Nonlinear Renewal Theory in Sequential Analysis

MICHAEL WOODROOFE
Department of Statistics
University of Michigan

SOCIETY for INDUSTRIAL and APPLIED MATHEMATICS • 1982

PHILADELPHIA, PENNSYLVANIA 19103

Library of Congress Catalog Card Number: 81-84856.
ISBN: 0-89871-180-0

Contents

Preface v

Introduction 1

Chapter 1
RANDOMLY STOPPED SEQUENCES
1.1. Stopping times 3
1.2. The optional stopping theorem 5
1.3. Anscombe's theorem 10

Chapter 2
RANDOM WALKS
2.1. The renewal theorem 15
2.2. First passage and residual waiting times 17
2.3. Spitzer's formula 20
2.4. On the asymptotic distribution of R_a 24

Chapter 3
THE SEQUENTIAL PROBABILITY RATIO TEST
3.1. Simple hypotheses 29
3.2. Composite hypotheses 35

Chapter 4
NONLINEAR RENEWAL THEORY
4.1. The stopping time and excess 41
4.2. The expected stopping time 46

Chapter 5
LOCAL LIMIT THEOREMS
5.1. Conditional probabilities 53
5.2. Densities 57

Chapter 6
OPEN-ENDED TESTS
6.1. In exponential families 61
6.2. Error probabilities 64
6.3. The expected sample size 68

Chapter 7
REPEATED SIGNIFICANCE TESTS
7.1. Likelihood ratio tests 71
7.2. Error probabilities 74
7.3. The expected sample size 81
7.4. The normal case 84

Chapter 8
MULTIPARAMETER PROBLEMS
8.1. Multiparameter exponential families 89
8.2. Repeated likelihood ratio tests 91
8.3. Examples 94

Chapter 9
ESTIMATION FOLLOWING SEQUENTIAL TESTING
9.1. The bias 99
9.2. Confidence intervals 100

Chapter 10
SEQUENTIAL ESTIMATION
10.1. Point estimation 105
10.2. Fixed width confidence intervals 110

Appendix
PROOF OF THE RENEWAL THEOREM 113

References 117

Preface

This work derives from a series of lectures which I gave during a Regional Conference in Statistics at Oklahoma State University in July of 1980. The Conference was organized by Professor Nitus Mukhopadhyay and sponsored by Oklahoma State University and the Conference Board of the Mathematical Sciences with the support of the National Science Foundation. I thank all for making the Conference and this book possible.

The subject of the book is nonlinear renewal theory and its applications to the sequential design of experiments, a relatively new development in probability theory and mathematical statistics. I attempt to describe the major developments of the theory and its principal applications as of July, 1980.

The book is intended primarily for researchers in probability and statistics, but is intended to be accessible to advanced graduate students, too. In writing, I have assumed that the reader is familiar with probability theory at the level of Chung (1977) and mathematical statistics at the level of Bickel and Doksum (1977). In addition, results from Chapters 2 and 3 of Lehmann (1959) are used in a few places. The most relevant aspects of probability theory are reviewed in Chapters 1 and 2 of the present work.

I want to thank the participants at the Conference, especially David Siegmund and Steven Lalley, for helpful suggestions and insightful comments. In particular, Chapter 9 is based on a talk given by Professor Siegmund. I also want to thank Professor Tony Tai for his careful reading of the manuscript and helpful suggestions.

Introduction

The purpose of this book is to describe a class of sequential tests, called repeated significance tests, and to develop the mathematical techniques for determining their properties. Many of the issues are illustrated by the phenomenon of sampling to a foregone conclusion, described in the following paragraphs.

Let $X_1, X_2, \cdots$ be independent normally distributed random variables with common unknown mean μ, $-\infty < \mu < \infty$, and known variance $\sigma^2 > 0$, and consider the problem of testing the null hypothesis H_0: $\mu = 0$. If a sample of fixed (nonrandom) size n were taken, then outcomes for which the absolute value of $S_n = X_1 + \cdots + X_n$ exceeds $3\sigma\sqrt{n}$ would be regarded as strong evidence against the null hypothesis, according to classical statistical theory. On the other hand, if data arrive sequentially, and if S_n is computed for each $n \geqq 1$, then $|S_n|$ is certain to exceed $3\sigma\sqrt{n}$ for some n, even if H_0 is true, for the law of the iterated logarithm asserts that

$$(0.1) \qquad \limsup_{n\to\infty} \frac{S_n - n\mu}{\sigma\sqrt{2n \log\log n}} = 1 \quad \text{w.p. 1.}$$

In fact, (0.1) holds under the sole assumption that $0 < \sigma^2 < \infty$, without the normality assumption. An unscrupulous experimenter might exploit this observation by taking a sample of size

$$(0.2) \qquad t = \inf\{n \geqq 1: |S_n| > c\sigma\sqrt{n}\},$$

where $c \geqq 3$, and reporting it as a sample of fixed size. In this case, the experimenter is certain to observe $|S_t| > 3\sigma\sqrt{t}$ and, having reported t as a fixed sample size, could assert that H_0: $\mu = 0$ is untenable.

Taking a sample of size t consists of observing the process $X_1, X_2, \cdots$ until $\bar{X}_n = S_n/n$ is significantly different from zero, as judged by classical statistical criteria. Simple modifications of this technique have recently been proposed by Robbins (1970) and Armitage (1975), especially for use in sequential clinical trials. For concreteness, suppose that there is a maximum sample size N, and define t by (2). Then one modification takes $T = \min(t, N)$ observations and rejects H_0 if and only if $t \leqq N$: that is, the procedure rejects H_0: $\mu = 0$ if and only if $|S_n| > c\sigma\sqrt{n}$ for some $n \leqq N$. The probability of falsely rejecting H_0 is then

$$(0.3) \qquad \alpha^* = P_0\{|S_n| > c\sigma\sqrt{n}, \exists n \leqq N\},$$

which may be controlled by choice of c, as described in Chapter 7. In fact, developing techniques for approximating α^* and related quantities is one of the central objectives of this book.

To understand why repeated testing is appropriate for many clinical trials, suppose that X_i represents the difference between the response to a new experimental treatment and an existing one by the ith subject or pair of subjects in a clinical trial. Then H_0: $\mu = 0$ is the hypothesis that there is no difference in the average responses to the two treatments. Suppose now that one observes a large value of $\bar{X}_n = S_n/n$ early in an experiment which calls for treating N (pairs of) subjects. Then one suspects a large difference in the average responses, leading to the following ethical dilemma: on one hand, it may be judged unethical to continue the experiment, since continuing requires giving a treatment which is suspected to be inferior; on the other, if one terminates the experiment early, then the results might not meet the accepted standards for statistical evidence, and might not be believed by the scientific community. Repeated testing provides a solution to both problems. One may alleviate the ethical problems by terminating the experiment as soon as there is strong evidence that $\mu \neq 0$; and, by reporting α^* as the type I error probability, one may quantify the remaining uncertainty, thus meeting accepted standards for statistical evidence.

In addition, there is substantial theoretical work which recommends the use of repeated significance tests. The most relevant is that of Schwarz (1962), (1968), who showed that approximations to optimal Bayesian tests of separated hypotheses may be obtained by performing repeated likelihood ratio tests, as defined in Chapters 7 and 8. Questions of optimality are neglected here, since they have been addressed in Chernoff's (1972) monograph. Rather, the properties of the tests are of primary concern here.

The mathematical techniques used to obtain these properties originated with the martingale techniques of Wald (1947) and the renewal theorem of Blackwell (1948), (1953) and Erdös, Feller and Pollard (1949). The identities of Baxter (1958) and Spitzer (1960) play an important role, too. Recently, the martingale techniques have been extended by Robbins (1970) and Robbins and Siegmund (1970), and renewal theory has been extended by Woodroofe (1976a) and Lai and Siegmund (1977), (1979). These results have been applied to determine the properties of repeated significance tests by Woodroofe (1976b), (1978), (1979) and Siegmund (1977), (1978), (1980), among others.

CHAPTER 1

Randomly Stopped Sequences

1.1. Stopping times. Let $(\mathscr{X}, \mathscr{A})$ be a measurable space; let $\mathscr{A}_n$, $n \geqq 1$, be an increasing sequence of sub-sigma-algebras of $\mathscr{A}$; and let $\mathscr{A}_\infty$ be the smallest sigma-algebra containing all $\mathscr{A}_n$, $n \geqq 1$. Then an extended random variable t is said to be a *stopping time* (with respect to $\mathscr{A}_n$, $n \geqq 1$) if and only if

$$t = 1, 2, \cdots \text{ or } \infty \quad \text{and} \quad \{t = n\} \in \mathscr{A}_n \quad \text{for all } n \geqq 1.$$

If P is a probability measure on $\mathscr{A}$, then t is said to be *proper* with respect to P if and only if $t < \infty$ w.p. 1 (P). In many examples, n indexes time, $\mathscr{A}_n$ represents the information available at time n and t is the time at which one ceases to observe the process. Then the condition $\{t = n\} \in \mathscr{A}_n$, $n \geqq 1$, requires that the decision to cease observing the process at a given time depends only on the information available then; t is proper if and only if one is certain to cease at some finite time. If t is a stopping time, then the events

$$\{t \leqq n\} = \bigcup_{k=1}^{n} \{t = k\} \quad \text{and} \quad \{t > n\} = \{t \leqq n\}'$$

are both in $\mathscr{A}_n$ for all $n \geqq 1$. In fact, t is a stopping time if and only if these events are in $\mathscr{A}_n$ for all $n \geqq 1$.

If t is a stopping time, then an event $A \in \mathscr{A}_\infty$ is determined prior to time t if and only if

$$A \cap \{t = n\} \in \mathscr{A}_n \quad \text{for all } n \geqq 1.$$

The class of all such events is easily seen to be a sigma-algebra, and is denoted by $\mathscr{A}_t$. If $t = m$ is a constant, then $\mathscr{A}_t = \mathscr{A}_m$, so the new notation is consistent with the old.

LEMMA 1.1. *If s and t are stopping times for which $s \leqq t$, then $\mathscr{A}_s \subset \mathscr{A}_t$.*

Proof. If $A \in \mathscr{A}_s$, then

$$A \cap \{t = n\} = \bigcup_{k=1}^{n} A\{t = n, s = k\} = \left[\bigcup_{k=1}^{n} A\{s = k\}\right] \cap \{t = n\} \tag{1.1}$$

for all $n \geqq 1$, since $s \leqq t$. For $k \leqq n$, $A\{s = k\} \in \mathscr{A}_k \subset \mathscr{A}_n$, so the event on the right side of (1.1) is in $\mathscr{A}_n$ for each $n \geqq 1$.

Example 1.1. A sequence of random variables $X_1, X_2, \cdots$ is said to be *adapted* to $\mathscr{A}_n$, $n \geqq 1$, if and only if X_n is $\mathscr{A}_n$-measurable for all $n \geqq 1$. If X_n, $n \geqq 1$, are adapted to $\mathscr{A}_n$, $n \geqq 1$, and if B_n, $n \geqq 1$, are Borel subsets of $(-\infty, \infty)$, then

$$t = \inf\{n \geqq 1 : X_n \in B_n\}$$

defines a stopping time, for $\{t = n\} = \{X_n \in B_n \text{ and } X_k \notin B_k, k < n\} \in \mathscr{A}_n$ for all $n \geqq 1$.

If X_n are adapted to $\mathcal{A}_n$, $n \geqq 1$, and t is a stopping time, then

$$X_t = \begin{cases} X_n & \text{if } t = n, \\ \limsup_{k\to\infty} X_k & \text{if } t = \infty \end{cases}$$

defines an extended, $\mathcal{A}_t$-measurable random variable; for if B is a Borel set of $(-\infty, \infty)$, then $\{t = n, X_t \in B\} = \{t = n, X_n \in B\} \in \mathcal{A}_n$ for all $n \geqq 1$.

Example 1.2. If P is a probability measure on $\mathcal{A}$, then random variables X_n, $n \geqq 1$, are said to be *independently adapted* to $\mathcal{A}_n$, $n \geqq 1$ (with respect to P) if and only if X_n, $n \geqq 1$, are adapted to $\mathcal{A}_n$, $n \geqq 1$, and $\mathcal{A}_n$ is independent of the sequence X_k, $k > n$, for every $n \geqq 1$. Then $X_1, X_2, \cdots$ must be independent (with respect to P). Suppose that $X_1, X_2, \cdots$ are independent and identically distributed (i.i.d.) with common distribution F and independently adapted to $\mathcal{A}_n$, $n \geqq 1$. If t is any stopping time, then X_{t+k}, $k \geqq 1$, are conditionally i.i.d. with common distribution F, given $\mathcal{A}_t$ on $\{t < \infty\}$; for, if $A \in \mathcal{A}_t$, then

$$\begin{aligned} P\{t<\infty, A, X_{t+1} \in B_1, \cdots, X_{t+k} \in B_k\} \\ &= \sum_{n=1}^{\infty} P\{t = n, A, X_{n+1} \in B_1, \cdots, X_{n+k} \in B_k\} \\ &= \sum_{n=1}^{\infty} P\{t = n, A\} \prod_{i=1}^{k} F\{B_i\} = \int_{A,t<\infty} \prod_{i=1}^{k} F\{B_i\}\, dP \end{aligned}$$

for Borel sets $B_1, \cdots, B_k$ of $(-\infty, \infty)$ and $k \geqq 1$. That is, $\prod_{i=1}^{k} F\{B_i\}$ is (a version of) the conditional probability $P\{X_{t+1} \in B_1, \cdots, X_{t+k} \in B_k | \mathcal{A}_t\}$ on $\{t < \infty\}$.

If P is a probability measure on $(\mathcal{X}, \mathcal{A})$ and if t is a stopping time, then $P^t = P|_{\mathcal{A}_t}$ denotes the restriction of P to $\mathcal{A}_t$; in particular, P^n denotes the restriction of P to $\mathcal{A}_n$, $n \geqq 1$. If Q is a second probability measure for which Q^n is absolutely continuous ($\ll$) with respect to P^n for every $n \geqq 1$, then one may form the likelihood ratios

$$L_n = \frac{dQ^n}{dP^n}, \qquad n \geqq 1.$$

The first theorem asserts that the likelihood ratios are transformed naturally by optional stopping.

THEOREM 1.1. *Let P and Q be probability measures for which $Q^n \ll P^n$ for all $n \geqq 1$, and let t be a stopping time. Then*

$$Q\{A, t<\infty\} = \int_{A,t<\infty} L_t\, dP \quad \text{for all } A \in \mathcal{A}_t.$$

If $t < \infty$ w.p. 1 (Q), then $Q^t \ll P^t$ and $dQ^t/dP^t = L_t I_{\{t<\infty\}}$, where I_A denotes the indicator of the event A.

Proof. If $A \in \mathcal{A}_t$, then $A \cap \{t = n\} \in \mathcal{A}_n$ for all $n \geqq 1$, so

$$Q\{A, t<\infty\} = \sum_{n=1}^{\infty} Q\{A, t = n\} = \sum_{n=1}^{\infty} \int_{A,t=n} L_n\, dP = \int_{A,t<\infty} L_t\, dP.$$

This establishes the first assertion, and the second follows easily.

COROLLARY 1.1. (the fundamental identity of sequential analysis). *If* $t<\infty$ *w.p.* 1 (Q), *then*

$$\int_{t<\infty} L_t\,dP=1.$$

In spite of their simplicity, the theorem and its corollary have important consequences in sequential analysis. When properly specialized, the theorem asserts that the likelihood function is unaffected by optional stopping.

Example 1.3. Let $X_1, X_2, \cdots$ be random variables on $(\mathscr{X}, \mathscr{A})$, and suppose that $\mathscr{A}_n=\sigma\{X_1, \cdots, X_n\}$, $n\geqq 1$. Let P_ω, $\omega\in\Omega$, be a parametric family of probability distributions over $(\mathscr{X}, \mathscr{A})$ and suppose that there is a probability measure λ for which $P^n \ll \lambda^n$ for all $\omega\in\Omega$ and $n\geqq 1$. If $n\geqq 1$, then (an appropriate version of) the likelihood ratios

$$L_n(\omega)=L_n(\omega; X_1, \cdots, X_n)=\frac{dP^n_\omega}{d\lambda^n}, \qquad \omega\in\Omega$$

is the likelihood function for the experiment in which $X_1, \cdots, X_n$ are observed. The theorem asserts that if t is any finite stopping time then $L_t(\omega)$ is the likelihood function for the experiment in which $X_1, \cdots, X_t$ are observed.

While the likelihood function is unaffected by optimal stopping, the sampling distributions of maximum likelihood estimators and likelihood ratio test statistics may be severely affected. This is illustrated by the phenomenon of sampling to a foregone conclusion, described in the Introduction.

Remarks and references. The book of Chow, Robbins, and Siegmund (1970) describes techniques for finding a stopping time which maximizes a given objective function. This is an interesting problem which is central to finding optimal sequential procedures.

The fact that likelihood functions are unaffected by optional stopping, but sampling distributions may be, has occasioned numerous articles in the literature. For different points of view, see the articles by Anscombe (1963), Armitage (1963), (1967), and Cornfield and Greenhouse (1967).

The fundamental identity was discovered by Wald (1947) in his study of the sequential probability ratio test.

1.2. The optional stopping theorem. Let $(\mathscr{X}, \mathscr{A}, P)$ be a probability space and let $\mathscr{A}_n$, $n\geqq 1$, be increasing sub-sigma-algebras of $\mathscr{A}$. A sequence Y_n, $n\geqq 1$, of random variables is said to be a *martingale* with respect to $\mathscr{A}_n$, $n\geqq 1$, if and only if Y_n, $n\geqq 1$, are adapted to $\mathscr{A}_n$, $n\geqq 1$; Y_n is integrable for $n\geqq 1$, and

$$E(Y_{n+1}|\mathscr{A}_n)=Y_n \quad \text{w.p. } 1, \quad n\geqq 1. \tag{1.2}$$

If $=$ is replaced by $\leqq$ in (1.2), then Y_n, $n\geqq 1$, is said to be a *supermartingale* with respect to $\mathscr{A}_n$, $n\geqq 1$; if $=$ is replaced by $\geqq$ in (1.2), then Y_n, $n\geqq 1$, is said to be a *submartingale* with respect to $\mathscr{A}_n$, $n\geqq 1$.

The following remark is useful in checking condition (1.2).

LEMMA 1.2. *Let $\mathscr{B}$ be a sub-sigma-algebra of $\mathscr{A}$, and let X and Y be integrable random variables for which X is $\mathscr{B}$-measurable. Then*

$$E(Y|\mathscr{B}) \leqq (=) X \quad w.p.\ 1 \tag{1.3}$$

if and only if

$$\int_B Y\,dP \leqq (=) \int_B X\,dP \quad \text{for all } B \in \mathscr{B}. \tag{1.4}$$

Proof. It suffices to prove the assertion for $\leqq$. Since

$$\int_B Y\,dP = \int_B E(Y|\mathscr{B})\,dP \quad \text{for all } B \in \mathscr{B},$$

by the definition of conditional expectation, it is clear that (1.3) implies (1.4). Conversely, if (1.4) holds for all $B \in \mathscr{B}$, then (1.4) holds when $B = \{E(Y|\mathscr{B}) > X\} = C$, say. In this case (1.4) asserts

$$\int [E(Y|\mathscr{B}) - X]^+\,dP = \int_C [E(Y|\mathscr{B}) - X]\,dP = \int_C (Y - X)\,dP \leqq 0,$$

which implies that $E(Y|\mathscr{B}) \leqq X$ w.p. 1.

It follows easily from the lemma that if Y_n, $n \geqq 1$, is a (sub, super) martingale with respect to any increasing sequence of sigma-algebras, then Y_n, $n \geqq 1$, is a (sub, super) martingale with respect to $\mathscr{A}_n = \sigma\{Y_1, \cdots, Y_n\}$, the sigma-algebra generated by $Y_1, \cdots, Y_n$, $n \geqq 1$. The unqualified term "(sub, super) martingale" means (sub, super) martingale with respect to $\sigma\{Y_1, \cdots, Y_n\}$, $n \geqq 1$.

Example 1.4. *Likelihood ratios.* Let $\mathscr{A}_n$, $n \geqq 1$, be any increasing sequence of sigma-algebras, let Q be a second probability measure on $\mathscr{A}$ and suppose that the restriction $Q^n = Q|_{\mathscr{A}_n}$ is absolutely continuous with respect to $P^n = P|_{\mathscr{A}_n}$ for $n \geqq 1$. Then the likelihood ratios

$$L_n = \frac{dQ^n}{dP^n}, \qquad n \geqq 1,$$

for a martingale with respect to $\mathscr{A}_n$, $n \geqq 1$, for if $A \in \mathscr{A}_n$ and $n \geqq 1$, then

$$\int_A L_{n+1}\,dP = Q^{n+1}(A) = Q^n(A) = \int_A L_n\,dP.$$

Example 1.5. *Sums of independent random variables.* Suppose that $X_1, X_2, \cdots$ are independently adapted to $\mathscr{A}_n$, $n \geqq 1$. If $E(X_k) = 0$, $k \geqq 1$, then the partial sums $S_n = X_1 + \cdots + X_n$, $n \geqq 1$, form a martingale with respect to $\mathscr{A}_n$, $n \geqq 1$; for S_n is $\mathscr{A}_n$-measurable, $E|S_n| < \infty$, and $E(S_{n+1}|\mathscr{A}_n) = E(S_n|\mathscr{A}_n) + E(X_{n+1}|\mathscr{A}_n) = S_n + E(X_{n+1}) = S_n$ for all $n \geqq 1$. If in addition X_k has a finite variance $\sigma_k^2 = E(X_k^2)$ for all $k \geqq 1$, then

$$Y_n = S_n^2 - \sum_{k=1}^n \sigma_k^2, \qquad n \geqq 1$$

form a martingale with respect to $\mathcal{A}_n$, $n \geqq 1$, for $E(S_{n+1}^2|\mathcal{A}_n) = S_n^2 + 2S_nE(X_{n+1}) + E(X_{n+1}^2) = S_n^2 + \sigma_{n+1}^2$, $n \geqq 1$.

The most striking feature of the martingale property (1.3) is that it is preserved by optional stopping.

THEOREM 1.2 (the optional stopping theorem). *Let Y_n, $n \geqq 1$, be a supermartingale with respect to an increasing sequence $\mathcal{A}_n$, $n \geqq 1$, and let s and t be two proper stopping times (with respect to $\mathcal{A}_n$, $n \geqq 1$) for which $s \leqq t$. If Y_s and Y_t are integrable and if*

$$\liminf_{n\to\infty} \int_{t>n} |Y_n|\, dP = 0, \tag{1.5}$$

then

$$E(Y_t|\mathcal{A}_s) \leqq Y_s \quad w.p.\ 1 \tag{1.6}$$

and if Y_n, $n \geqq 1$, is a martingale with respect to $\mathcal{A}_n$, $n \geqq 1$, then there is equality in (1.6).

Proof. It is shown that (1.4) holds with $\mathcal{B} = \mathcal{A}_s$, $X = Y_s$, and $Y = Y_t$. The key observation is that

$$(Y_t - Y_s) = \sum_{k=s+1}^{t} (Y_k - Y_{k-1}) = \sum_{k=1}^{\infty} (Y_k - Y_{k-1}) I_{\{s<k\leqq t\}}, \tag{1.7}$$

where I_A denotes the indicator of the event A. Suppose first that t is a bounded random variable, say $t \leqq n < \infty$ with probability 1, so that all summands with $k > n$ vanish in (1.7). If $A \in \mathcal{A}_s$, then $A\{s < k \leqq t\} = A\{s < k\} \cap \{t \geqq k\} \in \mathcal{A}_{k-1}$, $k \geqq 1$, so

$$\int_A (Y_t - Y_s)\, dP = \sum_{k=1}^{n} \int_{A, s<k\leqq t} E(Y_k - Y_{k-1}|\mathcal{A}_{k-1})\, dP \leqq 0, \tag{1.8}$$

which implies (1.6). In the general case, let n_k, $k \geqq 1$, be a subsequence along which the lim inf is attained in (1.5), and let $s_k = \min(n_k, s)$ and $t_k = \min(n_k, t)$ for $k \geqq 1$. Then s_k and t_k are stopping times and t_k is bounded for each $k \geqq 1$, so (1.6) holds with s and t replaced by s_k and t_k for each $k \geqq 1$. Moreover, if $A \in \mathcal{A}_s$, then $A\{s \leqq n_k\} \in \mathcal{A}_{s_k}$, so that (1.8) holds with s, t and A replaced by s_k, t_k and $A\{s \leqq n_k\}$ for each $k \geqq 1$. Now

$$\begin{aligned} \int_A (Y_t - Y_s)\, dP &= \int_A (Y_{t_k} - Y_{s_k})\, dP + \int_A [(Y_t - Y_{t_k}) - (Y_s - Y_{s_k})]\, dP, \\ \int_A (Y_{t_k} - Y_{s_k})\, dP &= \int_{A, s\leqq n_k} (Y_{t_k} - Y_{s_k})\, dP \leqq 0, \qquad k \geqq 1 \end{aligned} \tag{1.9}$$

by (1.8), since $\{s > n_k\} \subset \{s_k = n_k = t_k\}$, $k \geqq 1$. Thus, it suffices to show that the last integral in (1.9) approaches zero as $k \to \infty$. Since $\{s > n_k\} \subset \{t > n_k\}$, the absolute value of this integral is at most

$$\int_{t>n_k} [|Y_t| + |Y_{n_k}|]\, dP + \int_{s>n_k} [|Y_s| + |Y_{n_k}|]\, dP \leqq \int_{t>n_k} [|Y_t| + |Y_s| + 2|Y_{n_k}|]\, dP,$$

which approaches zero as $k \to \infty$ in view of (1.5) and the assumption that Y_s and Y_t are integrable. This completes the proof in the case of a supermartingale, and it is easy to check that there is equality in the case of a martingale.

COROLLARY 1.2. *Let Y_n, $n \geqq 1$, be a (sub, super) martingale with respect to $\mathcal{A}_n$, $n \geqq 1$. If t is a proper stopping time for which* (1.5) *holds and Y_t is integrable, then*

$$E(Y_t)\ (\geqq, \leqq) = E(Y_1).$$

Proof. This follows by applying Theorem 1.1 with $s = 1$.

Corollary 1.1 is a special case.

COROLLARY 1.3 (the submartingale inequality). *Let Y_n, $n \geqq 1$, be a submartingale, and let $M_n = \max\{Y_1, \cdots, Y_n\}$, $n \geqq 1$. Then*

$$P\{M_n > y\} \leqq \frac{1}{y}\int_{M_n > y} Y_n\, dP \leqq \frac{1}{y}E(Y_n^+), \qquad y > 1, \quad n \geqq 1. \tag{1.10}$$

Proof. Given $y > 1$ and $n \geqq 1$, let $s = \inf\{k \geqq 1\colon Y_k > y \text{ or } k \geqq n\}$. Then s is a stopping time (with respect to $\mathcal{A}_k = \sigma\{Y_1, \cdots, Y_k\}$, $k \geqq 1$) and $s \leqq n$, so Theorem 1.2 is applicable to s and $t = n$. Now $M_n > y$ if and only if $Y_s > y$ and $\{Y_s > y\} \in \mathcal{A}_s$, so

$$P\{Y_s > y\} \leqq \frac{1}{y}\int_{Y_s > y} Y_s\, dP \leqq \frac{1}{y}\int_{Y_s > y} Y_n\, dP$$

by Markov's inequality and the optional stopping theorem. This establishes the first inequality in (1.10); the second is obvious.

In applications of the submartingale inequality, the following lemma is often useful. Its proof follows directly from Jensen's inequality for conditional expectations.

LEMMA 1.3. *Let Y_n, $n \geqq 1$, be a (sub)martingale with respect to $\mathcal{A}_n$, $n \geqq 1$; let ϕ be an (increasing) convex function on $(-\infty, \infty)$; and let $Z_n = \phi(Y_n)$, $n \geqq 1$. If Z_n, $n \geqq 1$, are integrable, then Z_n, $n \geqq 1$, is a submartingale with respect to $\mathcal{A}_n$, $n \geqq 1$.*

Example 1.6. Let $X_1, X_2, \cdots$ be independent random variables for which $E(X_k) = 0$ and $E|X_k|^\alpha < \infty$ for $k \geqq 1$, where $\alpha > 1$. Then

$$P\{\max_{k \leqq n} |S_k| > y\} \leqq \frac{1}{y^\alpha}\int_{\max_{k \leqq n} |S_k| > y} |S_n|^\alpha\, dP \tag{1.11}$$

for all $y > 1$ and $n \geqq 1$. Indeed, this follows from applying Corollary 1.3 to $Y_n = |S_n|^\alpha$, $n \geqq 1$. When $\alpha = 2$, (1.11) is Kolmogorov's inequality.

THEOREM 1.3 (Wald's lemmas). *Let $X_1, X_2, \cdots$ be i.i.d. random variables which are independently adapted to increasing sigma-algebras $\mathcal{A}_n$, $n \geqq 1$, let $S_n = X_1 + \cdots + X_n$, $n \geqq 1$ and let t be a proper stopping time for which $E(t) < \infty$. If X_1 has a finite mean μ, then*

$$E(S_t) = \mu E(t); \tag{1.12}$$

and

(1.13) $$E[(S_t - t\mu)^2] = \sigma^2 E(t),$$

if X_1 has a finite variance σ^2 too.

Proof. There is no loss of generality in supposing that $\mu = 0$, in which case S_n, $n \geqq 1$, is a martingale with respect to $\mathcal{A}_n$, $n \geqq 1$. Thus, to prove (1.12), it suffices to verify the condition (1.5) with $Y_n = S_n$, $n \geqq 1$. Now, when $t > n$,

$$S_n \leqq \sum_{k=1}^{n} |X_k| \leqq \sum_{k=1}^{t} |X_k| = Z_t, \quad \text{say},$$

so it suffices to show that $E(Z_t) < \infty$. As in (1.7), one may write

$$Z_t = \sum_{k=1}^{\infty} |X_k| I_{\{t \geqq k\}},$$

so

$$E(Z_t) = \sum_{k=1}^{\infty} \int_{t \geqq k} |X_k| \, dP = E|X_1| \sum_{k=1}^{\infty} P\{t \geqq k\} = E|X_1| E(t) < \infty,$$

by the independence of X_k and $\{t \geqq k\}$ and the monotone convergence theorem. Equation (1.12) now follows from the optional stopping theorem.

If X_1 has mean 0 and finite variance σ^2, then $Y_n = S_n^2 - n\sigma^2$, $n \geqq 1$ is a martingale with respect to $\mathcal{A}_n$, $n \geqq 1$, by Example 1.5, so (1.13) holds for any bounded stopping time. For the extension to integrable t, let $t_n = \min(n, t)$ and $t'_n = \max(n, t)$, $n \geqq 1$. Then each t_n is a bounded stopping time and $t_n \to t$ w.p. 1, so that

$$E(S_t^2) \leqq \liminf_{n\to\infty} E(S_{t_n}^2) = \lim_{n\to\infty} \sigma^2 E(t_n) = \sigma^2 E(t),$$

by Fatou's lemma and the monotone convergence theorem. To establish the reverse inequality, it suffices to show that $E(S_t^2) \geqq E(S_{t_n}^2)$ for all $n \geqq 1$, since $E(S_{t_n}^2) = \sigma^2 E(t_n) \to \sigma^2 E(t)$, as above. Now,

$$\begin{aligned} E[S_t^2 - S_{t_n}^2] &= \int_{t>n} [2S_n(S_{t'_n} - S_n) + (S_{t'_n} - S_n)^2] \, dP \\ &\geqq \int_{t>n} 2S_n E[S_{t'_n} - S_n | \mathcal{A}_n] \, dP, \qquad n \geqq 1; \end{aligned}$$

and

$$E[S_{t'_n} - S_n | \mathcal{A}_n] = 0, \qquad n \geqq 1,$$

by the optional stopping theorem applied to $Y_k = S_k$, $s = n$, and t'_n. Indeed, the condition (1.5) was verified in the proof of (1.12).

Example 1.7. *Sampling to a foregone conclusion.* Let $X_1, X_2, \cdots$ be i.i.d. with unknown mean μ, $-\infty < \mu < \infty$, and known variance σ^2, $0 < \sigma^2 < \infty$, and consider the hypothesis H_0: $\mu = 0$. As noted in the Introduction, a dishonest experimenter might try to deceive the world by taking a sample of size

$$t_c = \inf\{n \geqq 1: |S_n| > c\sigma\sqrt{n}\},$$

where $c > 3$, and reporting that it was a sample of fixed (predetermined) size. This experimenter may succeed with the deception, but may expect to spend some time in the process, since $E(t_c) = \infty$ for all $c \geqq 1$. Indeed, if $E(t_c)$ were finite, then one would have the contradiction,

$$\sigma^2 E(t) = E(S_{t_c}^2) > c^2\sigma^2 E(t_c).$$

Remarks and references. Wald's lemma (1.12) appeared in Wald's (1947) development of the sequential probability ratio test and did much to help stimulate the subsequent development of martingales. The treatment given here follows Chow, Robbins and Teicher (1965). Example 1.7 was discovered by Blackwell and Freedman (1964) for Bernoulli variables and extended by Chow, Robbins and Teicher.

The brief introduction to martingale theory given here omits all reference to the martingale convergence theorem. For a more detailed account, see Chow, Robbins and Siegmund (1970, Chapt. 2).

1.3. Anscombe's theorem. In this section $X_1, X_2, \cdots$ denote i.i.d. random variables with finite mean μ and finite, positive variance σ^2, and $S_n = X_1 + \cdots + X_n$, $n \geqq 1$, denote the partial sums. The central limit theorem asserts that

$$S_n^* = \frac{S_n - n\mu}{\sigma\sqrt{n}}$$

converges in distribution to a standard normal random variable Z as $n \to \infty$: in symbols, $S_n^* \Rightarrow Z \sim N(0, 1)$. Anscombe's theorem asserts that this convergence persists when n is replaced by integer-valued random variables, t_a say, for which t_a/a converges in probability to a finite positive constant.

The most important idea in the proof of Anscombe's theorem is that of *uniform continuity in probability* (u.c.i.p.). A sequence Y_n, $n \geqq 1$, of random variables is said to be u.c.i.p. if and only if for every $\varepsilon > 0$ there is a $\delta > 0$ for which

$$(1.14) \qquad P\left\{\max_{0 \leqq k \leqq n\delta} |Y_{n+k} - Y_n| \geqq \varepsilon\right\} < \varepsilon \quad \text{for all } n \geqq 1.$$

Of course, if (1.14) holds for all sufficiently large n for a given δ, then it holds for all $n \geqq 1$ with a possibly smaller δ.

In Lemma 1.4 below, Y_n, $n \geqq 1$, are said to be *stochastically bounded* if and only if for every $\varepsilon > 0$ there is a $C > 0$ for which

$$P\{|Y_n| > C\} < \varepsilon \quad \text{for all } n \geqq 1.$$

In particular, if Y_n converges in distribution, then Y_n, $n \geqq 1$, are stochastically bounded.

LEMMA 1.4. *If Y_n, $n \geqq 1$, and Z_n, $n \geqq 1$, are u.c.i.p., then so is $Y_n + Z_n$, $n \geqq 1$. If in addition Y_n, $n \geqq 1$, and Z_n, $n \geqq 1$, are stochastically bounded, and if ϕ is any continuous function on R^2, then $\phi(Y_n, Z_n)$, $n \geqq 1$, is u.c.i.p.*

Proof. The first assertion is obvious, so only the second is proved. If $\varepsilon > 0$, then there is a $C > 0$ for which $P\{|Y_n| > C \text{ or } |Z_n| > C\} \leqq \varepsilon/3$ for all $n \geqq 1$. If ϕ is continuous, then there is an $\varepsilon' < \varepsilon/3$ for which $|\phi(y', z') - \phi(y, z)| < \varepsilon$ whenever $|y' - y| \leqq \varepsilon'$, $|y| \leqq C$, $|z' - z| \leqq \varepsilon'$ and $|z| \leqq C$. Thus, if $\delta > 0$, then $\max_{0 \leqq k \leqq n\delta} |\phi(Y_{n+k}, Z_{n+k}) - \phi(Y_n, Z_n)| \geqq \varepsilon$ implies

$$|Y_n| > C \quad \text{or} \quad |Z_n| > C \quad \text{or} \quad \max_{k \leqq n\delta} |Y_{n+k} - Y_n| \geqq \varepsilon' \quad \text{or}$$
$$\max_{k \leqq n\delta} |Z_{n+k} - Z_n| \geqq \varepsilon',$$

and the latter event has probability at most ε for all $n \geqq 1$ for sufficiently small $\delta > 0$.

Example 1.8. *Normalized partial sums.* If $X_1, X_2, \cdots$ are i.i.d. with finite mean μ and finite positive variance σ^2, then $Y_n = S_n^*$, $n \geqq 1$, is u.c.i.p. In the verification, one may suppose that $\mu = 0$ and $\sigma = 1$. Then

$$|S_{n+k}^* - S_n^*| \leqq \frac{1}{\sqrt{n}} |S_{n+k} - S_n| + \left[1 - \sqrt{\frac{n}{n+k}}\right] |S_n^*| \quad \text{for } k, n \geqq 1.$$

If $\varepsilon, \delta > 0$ and $k \leqq n\delta$, then the second term on the right is bounded by $C(\delta)|S_n^*|$, where $C(\delta) = 1 - (1+\delta)^{-1/2}$ and

$$P\left\{C(\delta)|S_n^*| > \frac{\varepsilon}{2}\right\} \leqq P\left\{|S_n^*| > \frac{\varepsilon}{2C(\delta)}\right\},$$

which tends to zero as $\delta \to 0$ uniformly in $n \geqq 1$, since $|S_n^*|$, $n \geqq 1$, are stochastically bounded. For the first term, Kolmogorov's inequality yields

$$P\left\{\max_{k \leqq n\delta} |S_{n+k} - S_n| \geqq \frac{\varepsilon\sqrt{n}}{2}\right\} \leqq \left(\frac{4}{n\varepsilon^2}\right) n\delta = \frac{4\delta}{\varepsilon^2},$$

which is independent of $n \geqq 1$ and tends to zero as $\delta \to 0$. That S_n^*, $n \geqq 1$, is u.c.i.p. now follows easily.

THEOREM 1.4 (Anscombe's theorem). *Suppose that $Y_1, Y_2, \cdots$ are u.c.i.p.; let t_a, $a > 0$, be integer-valued random variables for which t_a/a converges to a finite, positive constant c in probability and let $N_a = [ac]$, $a > 0$. Then*

$$Y_{t_a} - Y_{N_a} \to 0 \quad \textit{in probability as } a \to \infty.$$

If in addition Y_n converges in distribution to a random variable Y, then $Y_{t_a} \Rightarrow Y$ as $a \to \infty$.

Proof. One may suppose that $t_a/a \to 1$ in probability as $a \to \infty$. Let $\varepsilon > 0$ and let $\delta < 1$ be as in (1.14). Then, for all large a,

$$P\{|Y_{t_a} - Y_{N_a}| > \varepsilon\} \leqq P\left\{|t_a - N_a| > \frac{\delta N_a}{4}\right\} + P\{\max_{|n - N_a| \leqq \frac{1}{4}\delta N_a} |Y_n - Y_{N_a}| > \varepsilon\}.$$

The second term is less than ε by (1.14), the first approaches zero as $a \to \infty$, since $t_a/a \to 1$ in probability as $a \to \infty$. The first assertion of the theorem follows, since $\varepsilon > 0$ was arbitrary, and the second is an immediate consequence of the first.

COROLLARY 1.4. *Suppose that $X_1, X_2, \cdots$ are i.i.d. with finite mean μ and finite, positive variance σ^2. If t_a, $a>0$, are positive integer-valued random variables for which $t_a/a \to c$, $0<c<\infty$, in probability as $a\to\infty$, then*

$$S_{t_a}^* \quad and \quad S_{t_a}^{\#} = \frac{S_{t_a} - t_a\mu}{\sigma\sqrt{ac}}$$

converge in distribution to a standard normal random variable Z as $a\to\infty$.

The convergence of the distribution of S_n^* to normality in the central limit theorem has the important property that the mean and variance of the limiting distribution are the same as the common mean and variance of S_n^*, $n \geqq 1$. Moreover, von Bahr's (1965) extension of the central limit theorem asserts: if $X_1, X_2, \cdots$ are i.i.d. with finite mean μ, finite, positive variance σ^2, and finite αth absolute moment $E|X_1|^\alpha < \infty$, where $\alpha > 2$, then

$$E|S_n^*|^\alpha \to 2^{\alpha/2}\frac{\Gamma(\frac{1}{2}+\alpha/2)}{\sqrt{\pi}}$$

the αth absolute moment of the standard normal distribution. Von Bahr's extension is proved by a careful analysis of the characteristic functions of S_n^*, $n \geqq 1$. The convergence of moments in Anscombe's theorem is examined next.

The reader may recall that random variables Y_n, $n \geqq 1$, are said to be *uniformly integrable* (u.i.) if and only if

$$\lim_{y\to\infty} \sup_{n\geqq 1} \int_{|Y_n|>y} |Y_n|\, dP = 0;$$

if Y_n are u.i. and Y_n converge in distribution to a random variable Y, then $E|Y|<\infty$ and $E(Y_n)\to E(Y)$; and, if $Y_n \geqq 0$ and $Y_n \Rightarrow Y$, then $E(Y_n)\to E(Y)<\infty$ if and only if Y_n, $n\geqq 1$, are u.i. See, for example, Loève (1963, p. 183).

COROLLARY 1.5. *Let $X_1, X_2, \cdots$ be i.i.d. with finite mean μ and finite, positive variance σ^2; suppose that $X_1, X_2, \cdots$ are independently adapted to increasing sigma-algebras $\mathcal{A}_n$, $n\geqq 1$; let t_a, $a>0$, be stopping times with respect to $\mathcal{A}_n$, $n \geqq 1$; and define $S_{t_a}^{\#}$ as in Corollary 1.4. If $t_a/a \to c$ in probability as $a\to\infty$, where $0<c<\infty$, and if t_a/a, $a>0$, are u.i., then $(S_{t_a}^{\#})^2$, $a>0$, are u.i. and*

$$E|S_{t_a}^{\#}|^\alpha \to 2^{\alpha/2}\frac{\Gamma(\frac{1}{2}+\alpha/2)}{\sqrt{\pi}} \quad as\ a\to\infty \quad for\ 0<\alpha\leqq 2.$$

Proof. Since t_a/a, $a>0$, are u.i., $E(t_a)/a \to c$ as $a\to\infty$. Thus, by Wald's lemma,

$$E|S_{t_a}^{\#}|^2 = \frac{1}{ac\sigma^2}E[(S_{t_a} - t_a\mu)^2] = \frac{1}{ac\sigma^2}\sigma^2 E(t_a) \to 1$$

as $a\to\infty$. Since the asymptotic distribution of $S_{t_a}^{\#}$ is standard normal, it follows that $|S_{t_a}^{\#}|^2$, $a>0$, are uniformly integrable. The corollary then follows, since all lower powers of $|S_{t_a}^{\#}|$, $a>0$, must be u.i.

The final theorem gives some additional information about uniform integrability.

THEOREM 1.5. *Let Y_n, $n \geqq 1$, be random variables and let*

$$G(y) = \sup_{n \geqq 1} P\{|Y_n| > y\}, \qquad y > 0.$$

If $r > 0$ and $y^{r-1}G(y)$ is integrable with respect to Lebesgue measure over $(0, \infty)$, *then $|Y_n|^r$, $n \geqq 1$, are u.i.*

Proof. A simple integration by parts shows that

$$\begin{aligned} \int_{|Y_n|>y} |Y_n|^r \, dP &= y^r P\{|Y_n| > y\} + \int_y^\infty r z^{r-1} P\{|Y_n| > z\} \, dz \\ &\leqq y^r G(y) + \int_y^\infty r z^{r-1} G(z) \, dz \end{aligned} \tag{1.15}$$

for all $y > 0$ and $n \geqq 1$, and the last line in (1.15) approaches zero as $y \to \infty$, if $z^{r-1}G(z)$ is integrable over $(0, \infty)$.

Remarks and references. Theorem 1.4 and Corollary 1.4 were established by Anscombe (1952). Corollary 1.4 may also be deduced from the weak convergence of random broken lines to Brownian motion; see Billingsley (1968, pp. 143–148). Corollary 1.5 is a special case of a theorem of Chow, Hsiung and Lai (1979).

CHAPTER 2

Random Walks

2.1. The renewal theorem. In this chapter $X_1, X_2, \cdots$ denote i.i.d. random variables and S_n, $n \geqq 0$, denote the partial sums, $S_0 = 0$ and $S_n = X_1 + \cdots + X_n$, $n \geqq 1$. The common distribution of X_i, $i \geqq 1$, is denoted by F, and the mean and variance of F, if any, are denoted by μ and σ^2. In some contexts, it is convenient to regard S_n, $n \geqq 0$, as a random walk which moves a distance X_n at each integral time n. Alternatively, if $X_1 \geqq 0$, it is sometimes convenient to regard S_n, $n \geqq 0$, as the times at which a hypothetical piece of equipment breaks and is repaired, or renewed, in which case X_i, $i \geqq 1$, are the times between repairs.

If J is a subinterval of $(-\infty, \infty)$, then the number of visits of S_n, $n \geqq 0$, to J is the random variable

$$N\{J\} = \sum_{n=0}^{\infty} I_{\{S_n \in J\}}.$$

Alternatively, $N\{J\}$ is the number of renewals which occur in the time interval J in the second interpretation. The expected values of $N\{J\}$,

$$U\{J\} = \sum_{n=0}^{\infty} P\{S_n \in J\}, \qquad J \subset (-\infty, \infty),$$

define a measure U on the Borel sets of $(-\infty, \infty)$, called the *renewal measure.* Observe that $U\{(-\infty, \infty)\} = \infty$, so that U is an infinite measure. It is seen below that if F has a positive mean μ, then $U\{J\}$ is finite for every finite interval J.

Example 2.1. If F is the exponential distribution with failure rate $\lambda > 0$, then S_n has density

$$f_n(x) = \frac{\lambda^n}{(n-1)!} x^{n-1} e^{-\lambda x}, \qquad x > 0, \quad n \geqq 1.$$

Observe that $f_1(x) + f_2(x) + \cdots = \lambda$ for all $x > 0$. Thus, if $I \subset (0, \infty)$, then

$$U\{I\} = \sum_{n=1}^{\infty} \int_I f_n(x)\, dx = \int_I \sum_{n=1}^{\infty} f_n(x)\, dx = \lambda\, m_0(I),$$

where m_0 denotes Lebesgue measure. That is, the expected number of renewals in I is the failure rate λ times the length of I.

The renewal theorem asserts that Example 2.1 is typical, at least for intervals of the form $a + I$, where a is large. Unfortunately, the renewal theorem is complicated by the necessity of considering certain discrete cases separately.

A distribution F on the Borel sets of $(-\infty, \infty)$ is said to be *arithmetic* if and only if there is a $d>0$ for which

$$F\{0, \pm d, \pm 2d, \cdots\}=1.$$

Then there is a largest such d which is called the (arithmetic) *span* of F. If F is arithmetic with span d, then each S_n is a multiple of d with probability, so the measure U is supported by the lattice $\{0, \pm d, \pm 2d, \cdots\}$ too.

THEOREM 2.1. (renewal theorem). *If F has mean μ, $0<\mu<\infty$, then $U\{I\}<\infty$ for every finite interval I. If F is nonarithmetic, then*

$$\lim U\{a+I\}=\begin{cases}\mu^{-1}m_0(I) & \text{as } a\to\infty,\\ 0 & \text{as } a\to-\infty,\end{cases} \tag{2.1}$$

for every finite interval I, where m_0 denotes Lebesgue measure. If F is arithmetic with span $d>0$, then (2.1) *holds as $a\to\pm\infty$ through multiples of d with m_0 replaced by*

$$m_d\{J\}=d\cdot \#\{k: kd\in J\}, \qquad J\subset(-\infty, \infty).$$

The proof of the renewal theorem is presented in the Appendix.

COROLLARY 2.1. *There is a constant $C=C_F$ for which $U\{(a, a+x]\}\leqq C(1+x)$ for all $x>0$ and all a, $-\infty<a<\infty$.*

Proof. The renewal theorem asserts that $U\{(a, a+1]\}$ remains bounded as $a\to\pm\infty$, so there are a_0, C_0 for which $U\{(a, a+1]\}\leqq C_0$ for all $|a|>a_0$ and, since $U\{(a, a+1]\}\leqq U\{(-a_0, a_0+1]\}$ for $|a|\leqq a_0$, there is a C_1 for which $U\{(a, a+1]\}\leqq C_1$ for all a, $-\infty<a<\infty$. Thus, letting $[x]$ denote the greatest integer which is less than or equal to x, we have

$$U\{(a, a+x]\}\leqq \sum_{k=0}^{[x]} U\{(a+k, a+k+1]\}\leqq C_1([x]+1)$$

for all $x>0$ and $-\infty<a<\infty$.

There is a useful integral form of the renewal theorem. It is described in the special case that $X_1\geqq 0$, in which case $U\{I\}=0$ for $I\subset(-\infty, 0)$. Let $\mathscr{C}_0$ be the class of all nonnegative, nonincreasing, right continuous functions h, defined on $[0, \infty)$, which are integrable with respect to Lebesgue measure on $[0, \infty)$. If $g=h_1-h_2$ is the difference of two functions $h_1, h_2\in\mathscr{C}_0$, and if $X_1\geqq 0$, then the convolution of g and U,

$$(g*U)(a)=\int_0^a g(a-x)U\{dx\}, \qquad a\geqq 0,$$

is well defined and has nice asymptotic properties as $a\to\infty$.

THEOREM 2.2. *Suppose that $X_1\geqq 0$ w.p. 1, and let $g=h_1-h_2$, where $h_1, h_2\in\mathscr{C}_0$. If F is nonarithmetic, then*

$$(g*U)(a)\to\frac{1}{\mu}\int_0^\infty g(y)\,dy \quad \text{as } a\to\infty. \tag{2.2}$$

If F is arithmetic with span d, then (2.2) *holds as $a \to \infty$ through multiples of d with m_0 replaced by m_d.*

Proof. There is no loss of generality in supposing that $g \in \mathscr{C}_0$. Then $xg(x) \to 0$ as $x \to \infty$, and an integration by parts yields

$$\int_0^\infty g(y)\,dy = \int_0^\infty yG\{dy\}, \tag{2.3}$$

where G is the finite measure defined by $G\{(a, b]\} = g(a) - g(b)$, $0 \leqq a < b < \infty$. In particular, the integral on the right side of (2.3) is finite. Next, another integration by parts shows that

$$(g * U)(a) = g(a)U\{[0, a]\} + \int_0^a U\{(a-x, a]\}G\{dx\}, \qquad a > 0.$$

By Corollary 2.1, there is a C for which $U\{(a-x, a]\} \leqq C(1+x)$ for all $x \geqq 0$ and $a \geqq 0$. Thus, $g(a)U\{[0, a]\} \to 0$ as $a \to \infty$ and $U\{(a-x, a]\}$, $a \leqq 1$, are dominated by an integrable function (G). In the nonarithmetic case, $U\{(a-x, a]\} \to \mu^{-1}x$ for all $x > 0$ as $a \to \infty$, so

$$(g * U)(a) \to \frac{1}{\mu}\int_0^\infty xG\{dx\} = \frac{1}{\mu}\int_0^\infty g(x)\,dx$$

by the dominated convergence theorem. The arithmetic case may be handled similarly.

2.2. First passage and residual waiting times. Let $S_n = X_1 + \cdots + X_n$, $n \geqq 1$, be a random walk and, for $a \geqq 0$, let

$$\tau_a = \inf\{n \geqq 1 : S_n > a\}$$

be the time at which the random walk first reaches the height a, or ∞ if no such time exists. Next, define R_a on $\{\tau_a < \infty\}$ by

$$R_a = S_{\tau_a} - a.$$

Thus, R_a is the excess of the random walk over the boundary a at the time which it first crosses a. Alternatively, if S_n, $n \geqq 1$, represents the times at which equipment is renewed, then R_a is the *residual waiting time* until the next renewal after time a. The main result of this section gives the asymptotic distribution of R_a as $a \to \infty$.

If the mean μ is positive, then $S_n \to \infty$ w.p.1 by the strong law of large numbers, so $\tau_a < \infty$ for all $a \geqq 0$ w.p. 1. In the next section, it is shown that $\tau_a < \infty$ for all $a \geqq 0$ w.p. 1 if $\mu = 0$, too.

LEMMA 2.1. *If F has mean μ, $0 < \mu < \infty$, then $E(\tau_a) < \infty$ for all $a \geqq 0$.*

Proof. Suppose first that X_1 is bounded above, say $X_1 \leqq b$ w.p. 1. Let $\tau_a \wedge n = \min(n, \tau_a)$ for $n \geqq 1$ and $a \geqq 0$. Then $S_{\tau_a \wedge n} \leqq a + b$ w.p. 1, so that

$$\mu E(\tau_a \wedge n) = E[S_{\tau_a \wedge n}] \leqq a + b$$

by Wald's lemma for $n \geqq 1$ and $a \geqq 0$. So $E(\tau_a) = \lim_{n \to \infty} E(\tau_a \wedge n) \leqq (a+b)/\mu$.

In the general case, let $X_i' = X_i'(b) = \min(b, X_i)$ for $i \geqq 1$ and $b > 0$, and let $\mu' = \mu'(b) = E(X_1')$. Then $\mu' \to \mu$ as $b \to \infty$, so there is a $b > 0$ for which $\mu' > 0$. With obvious conventions, one then has $S_n' \leqq S_n$, $n \geqq 1$, $\tau_a \leqq \tau_a'$ and $E(\tau_a') < \infty$ for $a \geqq 0$.

In particular, it follows that $E(S_{\tau_a}) = \mu E(\tau_a)$ for all $a \geqq 0$.

The (strict, ascending) *ladder epochs* σ_k, $k \geqq 1$, are defined by

$$\sigma_k = \inf\{n > \sigma_{k-1}: S_n > S_{\sigma_{k-1}}\},$$

where $\sigma_0 = 0$ and the infimum of the empty set is ∞. Thus σ_k, $k \geqq 1$, are the times, if any, at which the random walk reaches new maxima. Observe that the first ladder epoch is $\sigma_1 = \tau_0$. It plays an important role below, and is denoted by τ. The (strict, ascending) *ladder heights* are the values $S_k^{\#} = S_{\sigma_k}$, $k \geqq 1$. It is easily seen that $(\sigma_k - \sigma_{k-1}, S_k^{\#} - S_{k-1}^{\#})$, $k \geqq 1$, are i.i.d. as (τ, S_τ).

THEOREM 2.3. *Suppose that* $0 < \mu < \infty$. *If* F *is nonarithmetic, then* R_a *has a limiting distribution* H *as* $a \to \infty$, *where*

$$H\{dr\} = \frac{1}{E(S_\tau)} P\{S_\tau > r\}\, dr, \qquad r \geqq 0. \tag{2.4}$$

If F *is arithmetic with span* $d > 0$, *then* R_a *has a limiting distribution* H_d *as* $a \to \infty$ *through multiples of* d. *The limiting distribution assigns masses*

$$H_d\{kd\} = \frac{d}{E(S_\tau)} P\{S_\tau \geqq kd\}, \qquad k \geqq 1. \tag{2.5}$$

Proof. Suppose first that X_1 is a positive nonarithmetic variable, so that $\tau = 1$ and $S_\tau = X_1$. Let F_n denote the distribution of S_n, $n \geqq 0$, and let $U = F_0 + F_1 + \cdots$ denote the renewal measure. Then

$$\begin{aligned} P\{R_a > r\} &= \sum_{n=1}^{\infty} P\{S_{n-1} \leqq a, S_n > a + r\} \\ &= \sum_{n=1}^{\infty} \int_0^a [1 - F(a + r - y)] F_{n-1}\{dy\} = \int_0^a [1 - F(a + r - y)] U\{dy\} \end{aligned} \tag{2.6}$$

for $a, r \geqq 0$. This is of the form considered in the integral form of the renewal theorem with $g(y) = 1 - F(r + y)$, $y \geqq 0$. Clearly, $g \in \mathscr{C}_0$, so

$$\int_0^a [1 - F(a + r - y)] U\{dy\} \to \frac{1}{\mu} \int_0^\infty [1 - F(r + y)]\, dy = \frac{1}{\mu} \int_r^\infty [1 - F(y)]\, dy,$$

establishing the theorem when $X_1 \geqq 0$.

In the general, nonarithmetic case, let $S_k^{\#}$, $k \geqq 0$, denote the ladder heights and let $\tau_a^{\#}$ and $R_a^{\#}$, $a \geqq 0$, denote the first passage and residual waiting times for the ladder heights. Then $R_a = R_a^{\#}$, $a \geqq 0$, since each first passage time τ_a must be a ladder epoch; since $S_k^{\#}$, $k \geqq 1$, are sums of i.i.d. positive random variables, $R_a^{\#}$ has the limiting distribution (2.4).

The arithmetic case may be handled similarly.

Example 2.2. The distribution F is said to have an exponential right rail if and only if $1-F(x)=C\exp(-\beta x)$ for all $x \geqq 0$ for some positive constants C and β. Then

$$P\{\tau_a<\infty, R_a>r\}=P\{\tau_a<\infty\}e^{-\beta r}, \qquad a, r \geqq 0.$$

For

$$\begin{aligned} P\{\tau_a<\infty, R_a>r\} &= \sum_{n=1}^{\infty} P\{\tau_a \geqq n, S_n>a+r\} \\ &= \sum_{n=1}^{\infty} \int_{\tau_a \geqq n} C\exp[-\beta(a+r-S_{n-1})]\,dP \\ &= \exp(-\beta r) \sum_{n=1}^{\infty} \int_{\tau_a \geqq n} C\exp[-\beta(a-S_{n-1})]\,dP = b\,e^{-\beta r}, \end{aligned}$$

say, for $r \geqq 0$; letting $r=0$ shows that $b=P\{\tau_a<\infty\}$. Of course, if $\mu>0$, then $\tau_a<\infty$ for all $a \geqq 0$ w.p. 1, so that R_a has the exponential distribution with failure rate β for all $a \geqq 0$.

Observe that if S_τ has $p+1$ moments, where $p>0$, then the asymptotic distribution (2.4) or (2.5) has p moments. The next theorem develops conditions under which the moments of R_a converge to those of (2.4) or (2.5).

THEOREM 2.4. *Suppose that* $0<\mu<\infty$ *and that* $E[\max(0, X_1)^{p+1}]<\infty$, *where* $p>0$. *Then* R_a^p, $a \geqq 0$, *are uniformly integrable. In fact,*

$$\int_0^{\infty} r^{p-1} \sup_{a \geqq 0} P\{R_a>r\}\,dr<\infty.$$

Proof. Since $\tau_a \geqq n$ implies that $S_{n-1} \leqq a$,

$$\begin{aligned} P\{R_a>r\} &= \sum_{n=1}^{\infty} P\{\tau_a \geqq n, S_n>a+r\} \\ &\leqq \sum_{n=1}^{\infty} P\{S_{n-1} \leqq a, S_n>a+r\} = \int_{-\infty}^{a} [1-F[a+r-y)]U\{dy\}, \end{aligned} \tag{2.7}$$

as in Theorem 2.1 and Example 2.1. Next, by Corollary 2.1, there is a C for which $U\{(k, k+1]\} \leqq C$ for all k, $-\infty<k<\infty$; so the right side of (2.7) is at most

$$\begin{aligned} &\sum_{k \leqq a} [1-F(a+r-k-1)]U\{(k, k+1]\} \\ &\leqq \sum_{k \leqq a} C[1-F(a+r-k-1)] \\ &\leqq C\sum_{j=0}^{\infty} [1-F(r+j-1)] \leqq C\int_{r-2}^{\infty} [1-F(y)]\,dy. \end{aligned} \tag{2.8}$$

Finally, r^{p-1} times the right side of (2.8) is integrable over $(0, \infty)$, since $E[\max(0, X_1)^{p+1}]<\infty$.

COROLLARY 2.2. *Suppose that* $\mu > 0$, *that* $E[\max(0, X_1)^2] < \infty$, *and that* F *is nonarithmetic. Then*

$$E(R_a) \to \rho = \frac{E(S_\tau^2)}{2E(S_\tau)}$$

and

$$E(\tau_a) = \frac{1}{\mu}(a + \rho) + o(1) \quad \text{as } a \to \infty.$$

If $X_1 \geqq 0$,

$$U\{[0, a]\} = \frac{1}{\mu} a + \frac{\mu^2 + \sigma^2}{2\mu^2} + o(1) \quad \text{as } a \to \infty.$$

Proof. The first assertion follows directly from the theorem, with $p = 1$. The second then follows from Wald's lemma, since

$$\mu E(\tau_a) = E(S_{\tau_a}) = a + E(R_a), \qquad a \geqq 0.$$

Finally, if $X_1 \geqq 0$, then $\rho = (\mu^2 + \sigma^2)/2\mu$ and $E(\tau_a) = U\{[0, a]\}$, $a \geqq 0$, so the final assertion follows from the second.

In the arithmetic case, the assertions of Corollary 2.2 hold as $a \to \infty$ through multiples of the span d, with $\rho = E[S_\tau(S_\tau + d)]/2E(S_\tau)$. The reader may supply the details.

Remarks and references. The renewal theorem was discovered by Blackwell (1948), (1953) and by Erdös, Feller and Pollard (1949). See Smith (1958) for an excellent review paper. Stone (1965a, b) has given asymptotic expansions for the renewal measure in the presence of higher moments: if X_1 has $p+1$ moments, where $p > 1$, then the final assertion of Corollary 2.2 holds with $o(1)$ replaced by $o[(1/a)^{p-1}]$, without the assumption that $X_1 \geqq 0$. Lorden (1970) has developed some interesting inequalities for the moments of R_a; in particular, $E(R_a) \leqq E(X_1^2)/E(X_1)$ for all $a > 0$.

2.3. Spitzer's formula. Theorem 2.3 shows that the asymptotic distribution of residual waiting time R_a is simply related to the distribution of the first (strict, ascending) ladder height S_τ, where

$$\tau = \inf\{n \geqq 1 : S_n > 0\}$$

denotes the first (strict ascending ladder) epoch. In this section the distribution of S_τ is related to the underlying distribution F of $X_1, X_2, \cdots$. It is convenient to relate the characteristic functions first. Thus, let

$$\phi(t)=\int_{-\infty}^{\infty} e^{itx}F\{dx\},$$

$$\beta(s,t)=\int_{\tau<\infty} s^{\tau} e^{itS_{\tau}}\,dP=\sum_{n=1}^{\infty} s^n \int_{\tau=n} e^{itS_n}\,dP,$$

$$\gamma(s,t)=\sum_{n=0}^{\infty} s^n \int_{\tau>n} e^{itS_n}\,dP$$

for $0\leqq|s|<1$ and $-\infty<t<\infty$.

LEMMA 2.2. $\beta+\gamma=1+s\gamma\phi$.

Proof. Denote the coefficients of s^n in the series defining $\beta(s,t)$ and $\gamma(s,t)$ by $\beta_n(t)$ and $\gamma_n(t)$ for $n\geqq 0$, so that $\beta(s,t)=s\beta_1(t)+s^2\beta_2(t)+\cdots$ and $\gamma(s,t)=\gamma_0(t)+s\gamma_1(t)+\cdots$. Then

$$(\beta_n+\gamma_n)(t)=\int_{\tau\geqq n} e^{itS_n}\,dP$$

$$=\left(\int e^{itX_n}\,dP\right)\left(\int_{\tau>n-1} e^{itS_{n-1}}\,dP\right)=\phi(t)\gamma_{n-1}(t)$$

for $-\infty<t<\infty$ and $n\geqq 1$, since X_n is independent of S_{n-1} and $\{\tau>n-1\}$ for all $n\geqq 1$. The lemma now follows by summing over $n\geqq 1$.

THEOREM 2.5 (Spitzer's formula). *For* $0\leqq|s|<1$ *and* $-\infty<t<\infty$,

$$\log\left(\frac{1}{1-\beta}\right)=\sum_{k=1}^{\infty}\frac{1}{k}s^k\int_{0+}^{\infty} e^{itx}F^{*k}\{dx\}, \tag{2.9}$$

where F^{*k} *denotes the k-fold convolution of F with itself and* log *denotes the principal branch of the complex logarithm.*

Proof. For sufficiently small s, $|\beta|<1$ and $|\gamma-1|<1$ for all t. For such s, the lemma may be rewritten $(1-\beta)/\gamma=1-s\phi$, so

$$\log\left(\frac{1}{1-\beta}\right)+\log\gamma=\log\left(\frac{1}{1-s\phi}\right)=\sum_{k=1}^{\infty}\frac{1}{k}s^k\phi^k \tag{2.10}$$

for $-\infty<t<\infty$ and sufficiently small $s>0$. For fixed s, the right side of (2.10) is the Fourier transform of the finite measure

$$F^s=\sum_{k=1}^{\infty}\frac{1}{k}s^kF^{*k},$$

since ϕ^k is the Fourier transform of F^{*k}, $k\geqq 1$. Similarly,

$$\log\left(\frac{1}{1-\beta}\right)=\sum_{k=1}^{\infty}\frac{1}{k}\beta^k=\text{Fourier transform of } B^s,$$

where

$$B^s = \sum_{k=1}^{\infty} \frac{1}{k} B_s^{*k}$$

with

$$B_s\{I\} = \int_{\tau<\infty, S_\tau \in I} s^\tau \, dP$$

for Borel sets $I \subset (-\infty, \infty)$. The series defining B^s converges because the total mass of B_s is $E(s^\tau) \leqq s < 1$. It is important to observe that B^s is supported by the interval $(0, \infty)$, in that $B^s\{I\} = 0$ for all $I \subset (-\infty, 0]$, since the same is true of B_s. Next, a similar argument shows that $\log \gamma$ is the Fourier transform of a finite (possibly signed) measure C^s for which $C^s\{I\} = 0$ for all $I \subset (0, \infty)$. Since the Fourier transform uniquely determines a measure, $B^s + C^s = F^s$ for sufficiently small $s > 0$; and since B^s and C^s are supported by different half axes,

$$B^s\{I\} = F^s\{I\} = \sum_{k=1}^{\infty} \frac{1}{k} s^k F^{*k}\{I\}, \qquad I \subset (0, \infty). \tag{2.11}$$

Equation (2.9) now follows by recalling that $\log(1/(1-\beta))$ is the Fourier transform of B^s and using (2.11) to compute the latter; the extension to all $|s| < 1$ may be accomplished by observing that both sides of (2.9) are analytic in s.

Next, let $\bar{\tau}$ denote the first weak descending ladder epoch,

$$\bar{\tau} = \inf\{n \geqq 1 : S_n \leqq 0\},$$

and let

$$\bar{\beta}(s, t) = \int_{\bar{\tau}<\infty} s^{\bar{\tau}} e^{itS_{\bar{\tau}}} \, dP$$

for $0 \leqq |s| < 1$ and $-\infty < t < \infty$.

COROLLARY 2.3.

$$\log\left(\frac{1}{1-\bar{\beta}}\right) = \sum_{k=1}^{\infty} \frac{1}{k} s^k \int_{-\infty}^{0} e^{itx} F^{*k}\{dx\} = \log \gamma$$

and

$$[1 - \bar{\beta}(s, t)][1 - \beta(s, t)] = 1 - s\phi(t). \tag{2.12}$$

Proof. The first equality follows by reversing the roles of the half axes in Theorem 2.5, and the second follows from observing that C^s is the restriction of F^s to $(-\infty, 0]$ in the proof of Theorem 2.5 and using this observation to compute $\log \gamma$, the Fourier transform of C^s. Equation (2.12) then follows from Lemma 2.2.

COROLLARY 2.4. *If F has mean μ, $0 < \mu \leqq \infty$, then*

$$E(\tau) = \frac{1}{P\{\bar{\tau} = \infty\}} = \exp\left\{\sum_{k=1}^{\infty} \frac{1}{k} P[S_k \leqq 0]\right\} < \infty. \tag{2.13}$$

Proof. Clearly, $P\{\bar{\tau}<\infty\}=\lim_{s\uparrow 1}\bar{\beta}(s,0)$, so

$$1/P\{\bar{\tau}=\infty\}=\lim_{s\uparrow 1} 1/(1-\bar{\beta}(s,0))=\lim_{s\uparrow 1}\gamma(s,0),$$

which is easily seen to be the exponential series in (2.13). Similarly, $E(\tau)=\lim_{s\uparrow 1}(1-\beta(s,0))/(1-s)$. So, setting $t=0$ in (2.12), dividing by $1-s$, and letting $s\uparrow 1$ yields $E(\tau)=1/P\{\bar{\tau}=\infty\}$. That $E(\tau)$ is finite when $\mu>0$ was shown in Lemma 2.1.

In particular, it follows from Corollary 2.4 that if $\mu>0$, then the series in (2.13) is finite. If the variance is finite, then more can be asserted.

COROLLARY 2.5. *If F has positive mean μ and finite variance σ^2, then*

$$U\{(-\infty,0]\}=\sum_{k=0}^{\infty}P\{S_k\leqq 0\}<\infty.$$

Proof. The expected value of $\frac{1}{2}\tau(\tau-1)$ is the limit of the derivative of $(1-\beta(s,0))/(1-s)=\gamma(s,0)$ as $s\uparrow 1$. This may be computed as in Corollary 2.4. After some algebra, one finds $E(\tau^2)/E(\tau)=2U\{(-\infty,0]\}-1$, finite or infinite. That $E(\tau^2)<\infty$ follows from Wald's lemma and the inequalities $\mu^2\tau^2\leqq 2S_\tau^2+2(S_\tau-\mu\tau)^2$ and $S_\tau^2\leqq X_\tau^2\leqq X_1^2+\cdots+X_\tau^2$.

Interest in Corollary 2.5 derives from two sources. First, it asserts that the renewal function $U(a)=U\{(-\infty,a]\}$ is finite for all a, $-\infty<a<\infty$, when $\mu>0$ and $\sigma^2<\infty$. Second, it is easy to convert Corollary 2.5 into the assertion: if F has finite mean μ and finite variance σ^2, then

$$\sum_{k=1}^{\infty}P\{|S_k-k\mu|>k\varepsilon\}<\infty \quad \text{for all } \varepsilon>0. \tag{2.14}$$

COROLLARY 2.6. *If F has mean $\mu=0$, then $\tau<\infty>\bar{\tau}$ w.p. 1, but $E(\tau)=\infty=E(\bar{\tau})$.*

Proof. Setting $t=0$ and letting $s\to 1$ in (2.12) shows that either $\beta(1-,0)=1$ or $\bar{\beta}(1-,0)=1$. That is, either $\tau<\infty$ w.p. 1 or $\bar{\tau}<\infty$ w.p. 1. Suppose that $\tau<\infty$ w.p. 1. Then $E(\tau)=1/P\{\bar{\tau}=\infty\}$, as in the proof of Corollary 2.4. If $E(\tau)$ were finite, then $E(S_\tau)=\mu E(\tau)=0$, which is false. Thus, $\bar{\tau}<\infty$ w.p. 1. The dual case may be handled similarly.

The most striking corollary to Theorem 2.5 relates the distributions B and $\bar{B}$ of S_τ and $S_{\bar{\tau}}$ to F. Here

$$B\{I\}=P\{(t<\infty,S_\tau\in I\} \quad\text{and}\quad \bar{B}\{I\}=P\{\bar{\tau}<\infty,S_{\bar{\tau}}\in I\}.$$

THE FACTORIZATION THEOREM. $F=B+\bar{B}-B*\bar{B}$.

Proof. Observe that β and $\bar{\beta}$ are convergent when $s=1$, in which case they are the characteristic functions of B and $\bar{B}$. Thus, setting $s=1$ in (2.12) yields $\phi(t)=\beta(1,t)+\bar{\beta}(1,t)-\beta(1,t)\bar{\beta}(1,t)$, $-\infty<t<\infty$, and the theorem follows from the unicity of characteristic functions.

Example 2.3. *Exponential tails.* If F has an exponential right tail, say $1-F(x)=C\exp(-\omega x)$ for all $x\geqq 0$ for some positive C and ω, then S_τ has a (possibly defective) exponential distribution, $P\{S_\tau>r\}=b\exp(-\omega r)$, $r>0$, where $b=$

$P\{\tau<\infty\}$. See Example 2.2. Then $\beta(1, t)=b\omega/(\omega-it)$, $-\infty<t<\infty$, and $\bar{\beta}$ and $\bar{B}$ may be found from (2.12).

Remarks and references. Theorem 2.5 was discovered by Baxter (1958) and Spitzer (1960); see Feller (1966, Chapts. 12, 18). Corollary 2.5 is a special case of a theorem of Baum and Katz (1965). See Lai (1979) for recent extensions.

2.4. On the asymptotic distribution of R_a. Let S_n, $n\geqq 0$, be a nonarithmetic random walk for which $0<\mu=E(S_1)<\infty$. Then the asymptotic distribution of residual waiting time has density

$$h(r)=\frac{1}{E(S_\tau)}P\{S_\tau>r\},\qquad r>0,$$

where τ denotes the first strict ascending ladder height; see § 2.2. In this section, Spitzer's formula is used to develop useful expressions for h and the corresponding distribution function H.

To begin, let $\mathcal{H}$ denote the Laplace transform of H,

$$\mathcal{H}(\alpha)=\int_0^\infty e^{-\alpha r}H\{dr\},\qquad \alpha\geqq 0.$$

Then

$$\mathcal{H}(\alpha)=\frac{E[1-e^{-\alpha S_\tau}]}{\alpha E(S_\tau)},\qquad \alpha>0,$$

by a simple integration by parts.

COROLLARY 2.7.

$$\mathcal{H}(\alpha)=\frac{1}{\alpha\mu}\exp\,[-B(\alpha)],\qquad \alpha>0,$$

where

$$B(\alpha)=\sum_{k=1}^\infty\frac{1}{k}E[e^{-\alpha S_k^+}],\qquad \alpha>0$$

and $^+$ *denotes positive part; if X_1 has a finite variance σ^2, then the mean of H is*

$$\rho=\frac{\mu^2+\sigma^2}{2\mu}-\sum_{k=1}^\infty\frac{1}{k}E(S_k^-),$$

where $^-$ *denotes negative part.*

Proof. Theorem 2.5 remains valid when *it* is replaced by $-\alpha$, where $\alpha>0$, in which case Theorem 2.5 asserts

$$E[1-e^{-\alpha S_\tau}]=\exp\left\{-\sum_{k=1}^\infty\frac{1}{k}\int_{S_k>0}e^{-\alpha S_k}\,dP\right\},\qquad \alpha>0$$

and

$$E(S_\tau)=\mu E(\tau)=\mu\,\exp\left\{\sum_{k=1}^\infty\frac{1}{k}P\{S_k\leqq 0\}\right\}$$

by Wald's lemma and Corollary 2.4. The first assertion now follows by substitution. The second follows similarly by differentiating twice.

Example 2.4. If X_1 has the normal distribution with mean $\mu > 0$ and variance $\sigma^2 > 0$, then S_n has the normal distribution with mean $n\mu$ and variance $n\sigma^2$ for $n \geqq 1$. In this case one finds that

$$E[e^{-\alpha S_k^+}] = \Phi\left(-\frac{\mu}{\sigma}\sqrt{k}\right) + \exp\left\{\left[\frac{\sigma^2\alpha^2}{2} - \mu\alpha\right]^k\right\}\Phi\left[-\left(\frac{\alpha\sigma^2 - \mu}{\sigma}\right)\sqrt{k}\right]$$

for $\alpha \geqq 0$ and $k \geqq 1$, where Φ denotes the standard normal distribution function. So, $B(\alpha)$ may be simply computed. Observe that $B(\alpha)$ assumes an especially simple form when $\alpha = 2\mu/\sigma^2$. Similarly, one finds that

$$\rho = \frac{\mu^2 + \sigma^2}{2\mu} - \sigma \sum_{k=1}^{\infty} \frac{1}{\sqrt{k}}\left[\Phi'\left(\frac{\mu\sqrt{k}}{\sigma}\right) - \left(\frac{\mu\sqrt{k}}{\sigma}\right)\Phi\left(-\frac{\mu\sqrt{k}}{\sigma}\right)\right].$$

The explicit calculations of Example 2.4 required a tractable expression for the distribution of S_n for each $n \geqq 1$. In the absence of such expressions, one may use Fourier analysis. Recall that the distribution F is said to be *strongly nonlattice* if and only if

$$\limsup_{|t|\to\infty} |\phi(t)| < 1,$$

where ϕ denotes the characteristic function of F. Then $\phi(t) \neq 1$ for all $t \neq 0$, so that one may form

$$\xi(t) = \log\left[\frac{1}{1-\phi(t)}\right], \qquad t \neq 0.$$

Here log denotes the principal branch of the complex logarithm.

THEOREM 2.6. *If F is strongly nonlattice, then*

$$\log \mathcal{H}(\alpha) = \frac{1}{\pi}\int_0^{\infty}\left(\frac{\alpha^2}{\alpha^2+s^2}\right)\frac{1}{s}\left[\mathcal{I}\xi(s) - \frac{\pi}{2}\right] ds$$
$$- \frac{1}{\pi}\int_0^{\infty}\left(\frac{\alpha}{\alpha^2+s^2}\right)[\mathcal{R}\xi(s) + \log(\mu s)]\, ds, \qquad \alpha > 0,$$

where $\mathcal{R}$ and $\mathcal{I}$ denote real and imaginary part. If, in addition, F has a finite variance σ^2, then

$$\rho = \frac{\mu^2+\sigma^2}{4\mu} + \frac{1}{\pi}\int_0^{\infty} s^{-2}[\mathcal{R}\xi(s) + \log(\mu s)]\, ds.$$

Proof. Suppose first that ϕ is integrable with respect to Lebesgue measure on $(-\infty, \infty)$, Then each S_k has a density f_k which may be computed by Fourier inversion as

$$f_k(x) = \frac{1}{2\pi}\int_{-\infty}^{\infty} e^{-itx}\phi(t)^k\, dt, \qquad -\infty < x < \infty.$$

So

$$\sum_{k=1}^{\infty} \frac{1}{k} \int_{S_k>0} e^{-\alpha S_k}\, dP = \sum_{k=1}^{\infty} \frac{1}{k} \int_0^{\infty} e^{-\alpha x} f_k(x)\, dx$$

$$= \sum_{k=1}^{\infty} \frac{1}{2\pi k} \int_0^{\infty} \int_{-\infty}^{\infty} e^{-(\alpha+it)x} \phi(t)^k\, dt dx$$

$$= \frac{1}{2\pi} \int_{-\infty}^{\infty} \int_0^{\infty} e^{-(\alpha+it)x} \xi(t)\, dx dt$$

$$= \frac{1}{\pi} \int_0^{\infty} \left(\frac{1}{\alpha+it}\right) \xi(t)\, dt, \qquad \alpha > 0.$$

Similarly,

$$\sum_{k=1}^{\infty} \frac{1}{k} \int_{-\infty}^{0} e^{\beta x} f_k(x)\, dx = \frac{1}{\pi} \int_0^{\infty} \left(\frac{1}{\beta - it}\right) \xi(t)\, dt, \qquad \beta > 0.$$

By adding these two expressions, observing that the imaginary parts vanish and letting $\beta \to 0$, one finds that

$$B(\alpha) = \frac{1}{\pi} \int_0^{\infty} \left(\frac{\alpha}{\alpha^2 + s^2}\right) [\mathcal{R}\xi(s) + \log(\mu s)]\, ds$$

$$-\frac{1}{\pi} \int_0^{\infty} \left(\frac{\alpha^2}{\alpha^2 + s^2}\right) \frac{1}{s} \left[\mathcal{I}\xi(s) - \frac{\pi}{2}\right] ds - \log(\mu\alpha).$$

after some simple analysis. The theorem then follows by substitution this expression for $B(\alpha)$ into Corollary 2.6 and differentiating to get ρ.

In the general case, one replaces X_k by $Y_k = X_k + \varepsilon Z_k$, $k \geqq 1$, where $\varepsilon > 0$ and Z_k, $k \geqq 1$, are i.i.d. standard normal random variables. Then Y_1 has an integrable characteristic function, so the special case is applicable to $Y_1, Y_2, \cdots$; and the general case may be deduced by letting $\varepsilon \downarrow 0$. The details of the limiting operation, while nontrivial, are omitted.

There is an alternative expression for the density h; and this expression is central to the development in Chapter 5.

THEOREM 2.7. *Let* $M = \min(S_1, S_2, \cdots)$. *Then*

$$h(r) = \frac{1}{\mu} P\{M > r\}, \qquad r > 0.$$

Proof. Let $t = \sup\{n \geqq 1: S_n = M\}$.
Then

$$P\{t = n, M > r\} = P\{S_k \geqq S_n, k < n, S_n > r\} P\{S_k - S_n > 0, k > n\}$$

$$= P\{S_k \leqq 0, k < n, S_n > r\} P\{\bar{\tau} = \infty\}$$

$$= P\{\tau = n, S_\tau > r\} \frac{1}{E(\tau)}, \qquad r > 0,$$

by simple symmetries of the random walk S_k, $k \geqq 1$, and Corollary 2.4. The theorem now follows by summing over $n \geqq 1$ and recalling that $E(S_\tau) = \mu E(\tau)$.

Remarks and references. Theorem 2.6 was given by Woodroofe (1979) under stronger assumptions. Steve Lalley (personal communication) showed me how to weaken these assumptions. The reader is invited to formulate a version of Theorem 2.6 which is valid in the arithmetic case.

CHAPTER 3

The Sequential Probability Ratio Test

3.1. Simple hypotheses. Let G_0 and G_1 be distinct, mutually absolutely continuous probability distributions, defined on a measurable space $(\mathcal{Y}, \mathcal{C})$, and let $Y_1, Y_2, \cdots$ be i.i.d. with common distribution G. The simple hypotheses $G = G_0$ versus $G = G_1$ are considered in this chapter. Let $\mathcal{A}_n = \sigma\{Y_1, \cdots, Y_n\}$ be the sigma-algebra generated by $Y_1, \cdots, Y_n$, let $\mathcal{A} = \sigma\{Y_1, Y_2, \cdots\}$ and let P_0 and P_1 be the unique probability measures on $\mathcal{A}$ under which $Y_1, Y_2, \cdots$ are i.i.d. with common distributions G_0 and G_1. Then the restrictions P_0^n and P_1^n of P_0 and P_1 to $\mathcal{A}_n$ are mutually absolutely continuous for every $n \geqq 1$. In fact, if

$$g(y) = \frac{dG_1}{dG_0}(y), \qquad y \in \mathcal{Y},$$

then

$$L_n = \prod_{i=1}^{n} g(Y_i)$$

is (a version of) the likelihood ratio dP_1^n/dP_0^n for every $n \geqq 1$. The sequential probability ratio test (S.P.R.T.) with boundaries A and $1/B$, where $A, B > 1$, continues sampling as long as $1/B \leqq L_n \leqq A$; it stops with the first n, if any, for which either $L_n < 1/B$ or $L_n > A$, and it rejects G_0 if and only if $L_n > A$ at the termination. That is, the S.P.R.T. takes

$$t = \inf\left\{n \geqq 1 \colon L_n < \frac{1}{B} \text{ or } L_n > A\right\}$$

observations and rejects G_0 if and only if $L_t > A$. It is shown below that $t < \infty$ w.p. 1 (P_0, P_1), so the test is well defined. Alternatively, the S.P.R.T. may be described in terms of the random walk

$$S_n = \log L_n = X_1 + \cdots + X_n, \qquad n \geqq 1,$$

where $X_i = \log g(Y_i)$, $i \geqq 1$, are i.i.d. under both P_0 and P_1. In fact, letting $a = \log A$ and $b = \log B$,

$$t = \inf\{n \geqq 1 \colon S_n < -b \text{ or } S_n > a\}, \tag{3.1}$$

and G_0 is rejected if and only if $S_t > a$.

LEMMA 3.1 (Stein's lemma). *Let $X_1, X_2, \cdots$ be i.i.d. random variables, defined on a probability space $(\mathcal{X}, \mathcal{A}, P)$, let $a, b > 0$ and define t by (3.1). If $P\{X_1 \neq 0\} > 0$, then $t < \infty$ w.p. 1. In fact, there are C and δ for which $0 < \delta < 1$ and $P\{t > n\} \leqq C\delta^n$ for all $n \geqq 1$.*

Proof. There is no loss of generality in supposing that $P\{X_1>0\}>0$, in which case there are x_0, $\varepsilon>0$ for which $P\{X_1>x_0\}>\varepsilon$. Let $c=a+b$ and let m be an integer for which $mx_0>c$. Then $P\{S_m>c\}>\varepsilon^m$, so that $\delta^m=P\{|S_m|\leqq c\}<1$. If $k\geqq 1$, then

$$\begin{aligned} P\{t>km\} &\leqq P\{-b\leqq S_{jm}\leqq a, j=1,\cdots,k\} \\ &\leqq P\{|S_{jm}-S_{(j-1)m}|\leqq c, j=1,\cdots,k\}=\delta^{km}. \end{aligned}$$

The lemma now follows, since $P\{t>n\}$ is decreasing in n.

In particular, Stein's lemma asserts that t has a finite expectation.

Now let α_0 and α_1 denote the error probabilities of S.P.R.T.,

$$\alpha_0=P_0\{L_t>A\} \quad \text{and} \quad \alpha_1=P_1\left\{L_t<\frac{1}{B}\right\}.$$

Then

$$\alpha_0=\int_{L_t>A}\left(\frac{1}{L_t}\right)dP_1=\int_{S_t>a}\exp(-S_t)\,dP_1$$

and

$$\alpha_1=\int_{L_t<1/B}L_t\,dP_0=\int_{S_t<-b}\exp(S_t)\,dP_0$$

by Theorem 1.1. Of course $1/L_t<1/A$ on $\{L_t>A\}$, so that $\alpha_0\leqq(1/A)P_1\{L_t>A\}=(1/A)(1-\alpha_1)$, and similarly $\alpha_1<(1/B)(1-\alpha_0)$. Wald (1947) argued that there should be approximate equality in these relations, since $1/B\leqq L_{t-1}\leqq A$ and a single Y_i should not have a large effect on the product. He then suggested solving the approximate equalities for α_0 and α_1 as

$$\alpha_0\approx\frac{B-1}{AB-1} \quad \text{and} \quad \alpha_1\approx\frac{A-1}{AB-1}. \tag{3.2}$$

Observe that $(B-1)/(AB-1)\sim 1/A$ and $(A-1)/(AB-1)\sim 1/B$ as $A, B\to\infty$. In fact, the approximations (3.2) are asymptotically incorrect by a constant factor, as shown below in Theorem 3.1.

Let F_0 and F_1 denote the distributions of $X_1=\log g(Y_1)$ under G_0 and G_1, and let μ_0 and μ_1 denote the means of F_0 and F_1. Then

$$-\infty\leqq\mu_0=\int\log g(y)G_0\{dy\}<0$$

and

$$0<\mu_1=\int\log g(y)G_1\{dy\}\leqq\infty,$$

by Jensen's inequality. In fact, $-\mu_0$ and μ_1 are the Kullback–Leibler information numbers for comparing the distributions G_0 and G_1. See, for example, Chernoff

(1972, pp. 48–50). Next, observe that F_0 and F_1 are mutually absolutely continuous, so one is arithmetic if and only if the other is, in which case the spans are equal, say $d>0$. Since $\mu_0<0<\mu_1$, $S_n \to -\infty$ w.p. 1 (P_0) and $S_n \to \infty$ w.p. 1 (P_1) as $n \to \infty$. Thus, if $\mu_1 - \mu_0 < \infty$, then the residual waiting times of the random walks $-S_n$ and S_n have limiting distributions under P_0 and P_1 (as $a \to \infty$ through multiples of d in the arithmetic case). Denote these asymptotic distributions by H_0 and H_1 and let

$$\gamma_i = \int_0^\infty e^{-r} H_i\{dr\}, \qquad i = 0, 1.$$

THEOREM 3.1. *Let $a = \log A$ and $b = \log B$ and suppose that $\mu_1 - \mu_0 < \infty$. Then*

$$\alpha_0 \sim \gamma_1 e^{-a} \quad \textit{and} \quad \alpha_1 \sim \gamma_0 e^{-b}$$

as $a, b \to \infty$ (through multiples of d in the arithmetic case).

Proof. Let $\tau_a = \inf\{n \geqq 1 : S_n > a\}$, $a \geqq 0$, denote the first passage times for S_n, $n \geqq 0$, and let $R_a = S_{\tau_a} - a$, on $\{\tau_a < \infty\}$, denote the residual waiting time. Then, since $S_t > a$ implies that $t = \tau_a$,

$$\alpha_0 = \int_{S_t > a} \exp(-S_t)\, dP_1 = e^{-a} \int_{S_t > a} \exp(-R_a)\, dP_1. \tag{3.3}$$

From above, $P_1\{S_t > a\} = 1 - \alpha_1 \to 1$ as $a, b \to \infty$. So the integral on the right side of (3.3) converges to γ_1. This establishes the first assertion; the second may be established similarly.

THEOREM 3.2. *Suppose that $E_i(X_1^2) < \infty$ and let ρ_i denote the mean of the asymptotic distribution H_i, $i = 0, 1$. Then*

$$E_0(t) = \frac{1}{|\mu_0|}\{b + \rho_0\} + o(1) \quad \textit{and} \quad E_1(t) = \frac{1}{\mu_1}\{a + \rho_1\} + o(1)$$

as $a, b \to \infty$ (through multiples of d in the arithmetic case) with $a e^{-b} \to 0 \leftarrow b e^{-a}$.

Proof. By Stein's lemma $E_1(t) < \infty$, so by Wald's lemma

$$\mu_1 E_1(t) = E_1(S_t) = \int_{S_t > a} S_t\, dP_1 + \int_{S_t < -b} S_t\, dP_1. \tag{3.4}$$

Now

$$\int_{S_t > a} S_t\, dP_1 = \int_{S_t > a} (a + R_a)\, dP_1 = a + \rho_1 + o(1),$$

since $a e^{-b} \to 0$ and R_a, $a \geqq 0$, are uniformly integrable. Moreover, for $b > 1$, the absolute value of the second integral in (3.4) is at most

$$\left| \int_{S_t < -b} S_t\, dP_1 \right| = \left| \int_{S_t < -b} S_t \exp(S_t)\, dP_0 \right| \leqq b e^{-b},$$

which tends to zero as $b \to \infty$. This establishes the second assertion; the first may be similarly established.

The constants γ_0, γ_1, ρ_0, and ρ_1 which appear in Theorems 3.1 and 3.2 are complicated, but may often be determined from the results of §§ 2.3 and 2.4. The following result exploits the fact that S_n, $n \geqq 1$, are log-likelihood ratios.

THEOREM 3.3. *Suppose that* $\mu_1 - \mu_0 < \infty$. *Then*

$$\gamma_0 = \frac{1}{|\mu_0|\Delta} \quad \textit{and} \quad \gamma_1 = \frac{1}{\mu_1 \Delta},$$

where

$$\Delta = \exp\left\{\sum_{k=1}^{\infty} \frac{1}{k}[P_0(S_k > 0) + P_1(S_k \leqq 0)]\right\}.$$

Proof. Let $\mathcal{H}_1$ denote the Laplace transform of H_1, so that $\gamma_1 = \mathcal{H}_1(1)$. Then integration by parts and Spitzer's formula show that

$$\mathcal{H}_1(\alpha) = \frac{1}{\alpha E_1(S_\tau)} E_1[1 - e^{-\alpha S_\tau}] = \frac{1}{\alpha E_1(S_\tau)} \exp\left\{-\sum_{k=1}^{\infty} \frac{1}{k} \int_{S_k > 0} e^{-\alpha S_k}\, dP_1\right\}$$

for $\alpha > 0$, where $\tau = \tau_0$ denotes the first strict ascending ladder epoch. Now, by Wald's lemma and Corollary 2.4,

$$E_1(S_\tau) = \mu_1 E_1(\tau) = \mu_1 \exp\left\{\sum_{k=1}^{\infty} \frac{1}{k} P_1(S_k \leqq 0)\right\}$$

and

$$\int_{S_k > 0} \exp(-S_k)\, dP_1 = P_0\{S_k > 0\}, \qquad k \geqq 1,$$

since S_k, $k \geqq 1$, are log-likelihood ratios. Thus, $\gamma_1 = 1/\mu_1 \Delta$.

Next, write $\Delta = \Delta(P_0, P_1)$. Then, reversing the roles of G_0 and G_1 shows that $\gamma_0 = 1/|\mu_0|\Delta(P_1, P_0)$. And $\Delta(P_1, P_0) = \Delta(P_0, P_1)$, since

$$P_1\{S_k = 0\} = \int_{S_k = 0} L_k\, dP_0 = P_0\{S_k = 0\}, \qquad k \geqq 1.$$

COROLLARY 3.1. $|\mu_0|\gamma_0 = \mu_1 \gamma_1$.

Example 3.1. *The normal case.* Suppose that G_0 is the standard normal distribution and that G_1 is the normal distribution with mean $\theta > 0$ and unit variance. Then

$$X_1 = \theta\left[Y_1 - \frac{\theta}{2}\right]$$

is normally distributed with variance θ^2 and means $-\theta^2/2$ and $\theta^2/2$, under G_0 and G_1. If we let Φ denote the standard normal distribution function, it follows that

$$P_1\{S_k \leqq 0\} = \Phi\left(-\frac{\theta}{2}\sqrt{k}\right) = P_0\{S_k > 0\}, \qquad k \geqq 1,$$

so

$$\log \Delta = 2 \sum_{k=1}^{\infty} \frac{1}{k} \Phi\left(-\frac{\theta}{2}\sqrt{k}\right).$$

The expression for $\log \Delta$ does not simplify, but is amenable to numerical calculation. Some typical values of $\log \Delta$ and $\gamma_0 = \gamma_1 = 2/\theta^2\Delta$ are given in Table 3.1. It is clear from Table 3.1 that Wald's approximations (3.2) may substantially overestimate the error probabilities, at least for large A and B.

The mean of the asymptotic distributions H_0 and H_1 may also be computed. One finds that $\rho_1 = -\rho_0 = \rho(\theta)$, where

$$\rho(\theta) = 1 = \frac{\theta^2}{4} - \theta \sum_{k=1}^{\infty} \left[\frac{1}{\sqrt{k}} \Phi'\left(\frac{\theta}{2}\sqrt{k}\right) - \frac{\theta}{2} \Phi\left(-\frac{\theta}{2}\sqrt{k}\right)\right].$$

Some typical values of $\rho(\theta)$ are included in Table 3.1 too.

Example 3.2. *The exponential case.* Suppose that G_0 is the standard exponential distribution and that G_1 is the exponential distribution with failure rate $\omega \neq 1$. Then

$$X_1 = (1-\omega) Y_1 + \log \omega,$$

$$\mu_0 = (1-\omega) + \log \omega \quad \text{and} \quad \mu_1 = \frac{1}{\omega}(1-\omega) + \log \omega.$$

TABLE 3.1
Values of $\gamma_0 = \gamma_1$ and $\rho_0 = \rho_1$ for the normal distribution with $\sigma^2 = 1$

θ	$\log \Delta$	$\gamma_0 = \gamma_1$	$\rho_0 = \rho_1 = \rho(\theta)$
.05	6.71359	.97144	.02958
.10	5.35650	.94348	.05959
.15	4.57477	.91633	.09024
.20	4.02851	.89004	.12160
.25	3.61133	.86451	.15443
.30	3.27578	.83972	.18631
.40	2.75855	.79230	.25370
.50	2.37031	.74762	.32385
.60	2.06361	.70553	.42068
.70	1.81313	.66589	.47262
.80	1.60375	.62857	.55138
.90	1.42570	.59344	.63313
1.00	1.27230	.56037	.71794
1.20	1.02169	.49998	.89697
1.40	.82644	.44654	1.08894
1.60	.67131	.39925	1.29434
1.80	.54640	.35743	1.51362
2.00	.44493	.32043	1.74726

The computations were done on an Apple II microcomputer, using formula (26.2.17) of Abramowitz and Stegun (1970) to compute the standard normal distribution function.

If $\omega < 1$, then X_1 has an exponential right tail, under G_1, since

$$P_1\{X_1 > x\} = P_1\left\{Y_1 > \frac{x - \log \omega}{1-\omega}\right\} = Ce^{-\beta x}, \qquad x > 0,$$

with $\beta = \omega/(1-\omega)$ and $C = \exp(\beta \log \omega)$. It follows that R_a has an exponential distribution with failure rate β for all $a \geqq 0$. In particular,

$$\gamma_1 = \mathscr{H}_1(1) = \frac{\beta}{1+\beta} = \omega,$$

and

$$\gamma_0 = \frac{\mu_1}{|\mu_0|}\gamma_1 = \frac{(1-\omega)+\omega \log \omega}{|(1-\omega)+\log \omega|}.$$

Similarly, if $\omega > 1$, then $-X_1$ has an exponential right tail under G_0: $P_0\{-X_1 > x\} = C\exp(-\beta x)$ for $x > 0$, where $\beta = 1/(\omega - 1)$ and $C = \exp(-\beta \log \omega)$. It then follows that

$$\gamma_0 = \frac{\beta}{1+\beta} = \frac{1}{\omega}$$

and

$$\gamma_1 = \frac{|\mu_0|}{\mu_1}\gamma_0 = \frac{(\omega-1)-\log \omega}{[\omega \log \omega - (\omega - 1)]}.$$

In the next section, there is interest in computing $P\{S_t > a\}$ and $E(t)$ when the common distribution of $Y_1, Y_2, \cdots$ is neither G_0 nor G_1. The following theorem provides a useful tool.

THEOREM 3.4. *Let G_0 and G_1 be mutually absolutely continuous probability distributions and let $g = dG_1/dG_0$. Let F be another probability distribution, and suppose that there is a $\kappa \neq 0$ for which*

$$\int g^\kappa \, dF = 1.$$

Let

$$G_0^\# = F \quad \text{and} \quad G_1^\#\{dy\} = g(y)^\kappa F\{dy\}, \qquad -\infty < y < \infty.$$

Then $G_1^\#$ is a probability distribution. Let $Y_1, Y_2, \cdots$ be i.i.d. with common distribution G. If $\kappa > 0$ (respectively, $\kappa < 0$), then the S.P.R.T. of $G = G_0$ versus $G = G_1$ with boundaries A and $1/B$, $A, B > 1$, is the same as the S.P.R.T. of $G = G_0^\#$ versus $G = G_1^\#$ with boundaries A^κ and $1/B^\kappa$ (respectively, $1/B^\kappa$ and A^κ).

Proof. Let $P_i^\#$ be the probability measure under which $Y_1, Y_2, \cdots$ are i.i.d. with common distribution $G = G_i^\#$, $i = 0, 1$. Then

$$\log \frac{dP_1^{\#n}}{dP_0^{\#n}} = \kappa \log g(Y_1) + \cdots + \kappa \log g(Y_n) = \kappa \log \frac{dP_1^n}{dP_0^n}, \qquad n \geqq 1.$$

The theorem follows easily.

For example, when $G=F$ and $\kappa>0$, one finds that

$$P\{S_t>a\}\sim\frac{\gamma_1^{\#}}{A^{\kappa}}\quad \text{as } A, B\to\infty,$$

where $\gamma_1^{\#}$ is computed as was γ_1, but with G_1 and G_0 replaced by $G_1^{\#}$ and $G_0^{\#}$. Specific examples appear in the next section.

Remarks and references. The formulation of the S.P.R.T., (3.2) and Theorem 3.4 are taken directly from Wald's (1947) influential monograph. See also Lehmann (1959, §§ 3.10–3.12).

Theorems 3.1 and 3.2 are taken from Siegmund (1975), who gives additional terms. In effect, Siegmund suggests combining renewal theory with Wald's derivation of (3.2) and arrives at the approximation

$$\alpha_0\approx\frac{\dfrac{\gamma_1}{A}-\dfrac{\gamma_0\gamma_1}{AB}}{1-\dfrac{\gamma_0\gamma_1}{AB}} \tag{3.5}$$

among others. See also Keener (1980).

Theorem 3.3 is implicit in Siegmund (1975) and explicit in Lorden (1977).

Formally setting $b=\infty$ in Theorem 3.4 yields $P\{\max(S_1, S_2, \cdots)>a\}\sim \gamma_1^{\#}e^{-\kappa a}$ as $b\to\infty$ when $\kappa>0$. This is Cramér's estimate for the probability of ruin; see Feller (1966, p. 363).

3.2. Composite hypotheses. The S.P.R.T. as developed in § 3.1 is a test of the simple hypotheses $G=G_0$ versus $G=G_1$. In this section it is shown that the S.P.R.T. may be used to test composite hypotheses too. For simplicity, the discussion is limited to one-parameter exponential families.

Let Ω be an interval and let G_ω, $\omega\in\Omega$, denote a one-parameter, exponential family of probability distributions with natural parameter space Ω. Thus

$$G_\omega\{dy\}=\exp[\omega y-\psi(\omega)]\lambda\{dy\},\qquad -\infty<y<\infty,\quad \omega\in\Omega,$$

for some nondegenerate, sigma-finite measure λ. Here the normalizing constant is written $\exp[\psi(\omega)]=\int e^{\omega y}\lambda\{dy\}$ and the natural parameter space is the set of all ω for which this integral is finite. It is known that the natural parameter space is an interval and that ψ is differentiable on its interior Ω^0. In fact, if $Y\sim G_\omega$ then the mean and variance of Y are

$$\theta=E_\omega(Y)=\psi'(\omega)\quad\text{and}\quad D_\omega(Y)=\psi''(\omega),$$

where $'$ denotes differentiation. Since λ is assumed to be nondegenerate, the variance $\psi''(\omega)$ is positive for $\omega\in\Omega^0$. It follows that the mean $\theta=\psi'(\omega)$ is increasing in ω, and therefore a one-to-one function. Thus the family G_ω, $\omega\in\Omega$, could be parameterized by θ as well as ω. See Lehmann (1959, § 2.7) for the elementary properties of exponential families, and Barndorff-Nielson (1978) for a more detailed account.

Examples 3.3. (i) The family of normal distributions with unknown mean θ, $-\infty<\theta<\infty$, and unit variance forms an exponential family with $\theta=\omega$.

(ii) The family of exponential distributions with unknown failure rate $|\omega|$, $-\infty<\omega<0$ forms an exponential family.

(iii) The family of Poisson distributions with unknown mean $\theta>0$ is an exponential family with $\omega=\log\theta$.

(iv) The family of Bernoulli distributions with unknown mean θ, $0<\theta<1$, is an exponential family with $\omega=\log[\theta/(1-\theta)]$.

Now let $Y_1, Y_2, \cdots$ be i.i.d. with common distribution G_ω, for some unknown $\omega\in\Omega$, and write $P=P_\omega$ to emphasize the dependence on ω. Thus, P_ω is the unique probability measure on $\mathscr{A}=\sigma\{Y_1, Y_2, \cdots\}$ under which $Y_1, Y_2, \cdots$ are i.i.d. with common distribution G_ω. Let P_ω^n be the restriction of P_ω to $\mathscr{A}_n=\sigma\{Y_1,\cdots,Y_n\}$, $n\geqq 1$, $\omega\in\Omega$. Then

$$\frac{dP^n_{\omega_1}}{dP^n_{\omega_0}}=\exp\{(\omega_1-\omega_0)n\bar{Y}_n-n[\psi(\omega_1)-\psi(\omega_0)]\}$$

for all $n\geqq 1$ and all $\omega_0, \omega_1\in\Omega$.

Now let $\omega_0\in\Omega^0$ and consider testing the hypothesis $\omega\leqq\omega_0$. Given $\omega_1>\omega_0$, one may construct a S.P.R.T. of $\omega=\omega_0$ versus $\omega=\omega_1$ and use it to test the composite hypothesis $\omega\leqq\omega_0$. Let

(3.6) $$S_n=(\omega_1-\omega_0)n\bar{Y}_n-n[\psi(\omega_1)-\psi(\omega_0)],\qquad n\geqq 1,$$

denote the log-likelihood ratios and observe that the Kullback–Leibler information measure is

$$J(\omega_0,\omega_1)=E_{\omega_1}(S_1)=(\omega_1-\omega_0)\psi'(\omega_1)-[\psi(\omega_1)-\psi(\omega_0)].$$

Next, let $A, B>1$; and let $a=\log A$ and $b=\log B$. Then the S.P.R.T. of $\omega=\omega_0$ versus $\omega=\omega_1$ with boundaries A and $1/B$ takes

$$t=\inf\{n\geqq 1: S_n<-b \text{ or } S_n>a\}$$

observations and rejects $\omega=\omega_0$ if and only if $S_t>a$. When regarded as a test of $\omega\leqq\omega_0$, the S.P.R.T. rejects $\omega\leqq\omega_0$ in favor of $\omega>\omega_0$ if and only if $S_t>a$. Let β denote the power function of the test,

$$\beta(\omega)=P_\omega\{S_t>a\},\qquad \omega\in\Omega.$$

It is shown below that β is a nondecreasing function of ω. Thus, the maximum probability of a type I error is $\alpha_0=\beta(\omega_0)$, which may be approximated from Theorem 3.1. For example, if the distribution of S_1 is nonarithmetic, then

(3.7) $$\beta(\omega_0)\sim\gamma_1 A^{-1}$$

and

$$E_{\omega_0}(t)=\frac{1}{J(\omega_1,\omega_0)}\{b+\rho_0\}+o(1),$$

as $a, b \to \infty$ with $a\,e^{-b} \to 0 \leftarrow b\,e^{-a}$, where γ_1 and ρ_0 are as in Theorems 3.1 and 3.2; that is, γ_1 is the Laplace transform of the asymptotic distribution of residual waiting time for the random walk S_n, $n \geqq 1$, of (3.6), computed under the assumption that $\omega = \omega_1$, and ρ_0 is the mean of the asymptotic distribution of residual waiting time for $-S_n$, $n \geqq 1$, computed under the assumption that $\omega = \omega_0$. Of course, similar approximations are valid when $\omega = \omega_1$, and related approximations hold in the arithmetic case, provided that $a, b \to \infty$ through multiples of the span.

Approximations to the power function and expected sample size when $\omega_0 \neq \omega \neq \omega_1$ may be obtained by applying Theorem 3.4 to

$$G_0 = G_{\omega_0}, \quad G_1 = G_{\omega_1}, \quad F = G_\omega.$$

To apply this theorem, one must solve the equation $E_\omega[\exp(\kappa S_1)] = 1$. The expectation may be computed in terms of the function ψ; and the equation may be written

$$\psi[\omega + \kappa(\omega_1 - \omega_0)] - \psi(\omega) = \kappa[\psi(\omega_1) - \psi(\omega_0)]. \tag{3.8}$$

Since ψ is convex, (3.8) may have at most one solution $\kappa \neq 0$. Suppose that (3.8) has a solution $\kappa > 0$; and let $\omega_0^\# = \omega$ and $\omega_1^\# = \omega + \kappa(\omega_1 - \omega_0)$. Then the S.P.R.T. of ω_0 versus ω_1 with boundaries A and $1/B$ is the same as the S.P.R.T. of $\omega_0^\#$ versus $\omega_1^\#$ with boundaries A^κ and $1/B^\kappa$. So, if S_1 has a nonarithmetic distribution,

$$\beta(\omega) \sim \gamma_1^\# A^{-\kappa} \tag{3.9}$$

and

$$E_\omega(t) = J(\omega_1^\#, \omega_0^\#)^{-1}\{\kappa b + \rho_0^\#\} + o(1),$$

as $a, b \to \infty$ with $a\,e^{-b} \to 0 \leftarrow b\,e^{-a}$, where $\gamma_1^\#$ and $\rho_0^\#$ are computed as were γ_1 and ρ_0, but with ω_0 and ω_1 replaced by $\omega_0^\#$ and $\omega_1^\#$. Similar approximations hold if (3.8) has a negative solution and related approximations hold in the arithmetic case, provided that $a, b \to \infty$ through multiples of the span.

The approximations (3.9) are recommended when $\beta(\omega)$ is small, say $\beta(\omega) \leqq 0.01$. For larger values of $\beta(\omega)$, approximations based on (3.5) may be much more accurate.

Example 3.4. Suppose that G_ω is the normal distribution with mean ω, $-\infty < \omega < \infty$, and unit variance; and suppose that $\omega_0 = -\delta$ and $\omega_1 = \delta$. Then $S_n = 2\delta n \bar{Y}_n$, $n \geqq 1$. So, $E_\omega[\exp(\kappa S_1)] = \exp(2\delta\kappa\omega + 2\delta^2\kappa^2) = 1$ if and only if $\kappa = 0$ or $\kappa = -\delta^{-1}\omega$.

The monotonicity of the power function is considered next. Recall that a distribution F_1 is said to be *stochastically less than or equal to* (s.l.e.) another distribution F_2 if and only if $F_2(x) \leqq F_1(x)$ for all x, $-\infty < x < \infty$. The following sufficient condition is needed.

LEMMA 3.2. *Suppose that $F_2 \ll F_1$ and that (some version of) the density $f = dF_2/dF_1$ is nondecreasing on $(-\infty, \infty)$. Then F_1 is s.l.e. F_2.*

Proof. It is shown that if w is any bounded nonincreasing function, then $\int w\, dF_2 \leqq \int w\, dF_1$. The lemma then follows by applying this result to $w = I_{(-\infty,x]}$ for arbitrary x, $-\infty < x < \infty$. If w is any bounded, nonincreasing function, then

$$\text{(3.10)} \qquad \int w\, dF_2 - \int w\, dF_1 = \int w[f-1]\, dF_1 = \int [w-c][f-1]\, dF_1$$

for any c, $-\infty < c < \infty$. There is no loss of generality in supposing that $f(x) > 1$ for some x; for, otherwise, $F_1 = F_2$. Then there is an x_0 for which $f(x) \leqq 1$ for all $x < x_0$ and $f(x) \geqq 1$ for all $x > x_0$. Let $c = w(x_0)$ in (10). Then the integrand $[w-c][f-1]$ is nonpositive, so the integral is nonpositive, too.

LEMMA 3.3. *Let U denote a random variable which is uniformly distributed over* (0, 1). *If F_1 is s.l.e. F_2, then there are functions w_1 and w_2 for which*

$$w_1(u) \leqq w_2(u) \quad \textit{for } 0 < u < 1,$$

$$w_1(U) \sim F_1 \quad \textit{and} \quad w_2(U) \sim F_2.$$

Proof. The functions $w_i(u) = \inf\{x : F_i(x) \geqq u\}$, $0 < u < 1$, $i = 1, 2$, have the desired properties.

Observe that if G_ω, $\omega \in \Omega$, is an exponential family, then G_ω have monotone likelihood ratio—that is, if $\omega_1 < \omega_2$, then

$$\frac{dG_{\omega_2}}{dG_{\omega_1}} = \exp\{(\omega_2 - \omega_1)y - [\psi(\omega_2) - \psi(\omega_1)]\}$$

is an increasing function of y. It follows that G_ω, $\omega \in \Omega$, are stochastically increasing (that is, G_{ω_1} is s.l.e. G_{ω_2} whenever $\omega_1 < \omega_2$).

In the next theorem, let $-\infty \leqq b_n \leqq a_n \leqq \infty$, $n \leqq 1$, let $Y_1, Y_2, \cdots$ be i.i.d. with common distribution G_ω under P_ω let

$$s_n = Y_1 + \cdots + Y_n, \qquad n \geqq 1,$$

let

$$\tau = \inf\{n \geqq 1 : s_n < b_n \text{ or } s_n > a_n\}$$

and let

$$\Delta(\omega) = P_\omega\{\tau < \infty, s_\tau > a_\tau\}.$$

THEOREM 3.5. *With the notation of the previous paragraph, $\Delta(\omega)$ is a nondecreasing function of $\omega \in \Omega$.*

Proof. Let $\omega_1 < \omega_2$. Then there are i.i.d. random variables $U_1, U_2, \cdots$ which are uniformly distributed over (0, 1) and functions w_1 and w_2 on (0, 1) for which $w_1 \leqq w_2$, $Y_{1i} = w_1(U_i) \sim G_{\omega_1}$, and $Y_{2i} = w_2(U_i) \sim G_{\omega_2}$, $i \geqq 1$. Let C denote the set of sequences $s = (s_1, s_2, \cdots)$ for which $s_k \geqq b_k$ for all $k < n$ and $s_n > a_n$ for some $n \geqq 1$. Then $\Delta(\omega_1)$ is the probability that $s_1 = (s_{11}, s_{22}, \cdots) \in C$; and $\Delta(\omega_2)$ is the probability that $s_2 = (s_{21}, s_{22}, \cdots) \in C$. Since $w_1 \leqq w_2$, the former event implies the latter, so $\Delta(\omega_1) \leqq \Delta(\omega_2)$. Since $\omega_1 < \omega_2$ were arbitrary, the theorem follows.

The power function of the S.P.R.T. of $\omega \leqq \omega_0$ is of the form Δ, since $t < \infty$ w.p. 1 (P_ω) for all $\omega \in \Omega$. Thus, the power function of the S.P.R.T. is an increasing function.

Remarks and references. The test which takes τ observations and rejects $\omega \leqq \omega_0$ if and only if $s_\tau > a_\tau$ is a generalized S.P.R.T. as defined by Kiefer and Weiss (1957).

CHAPTER 4

Nonlinear Renewal Theory

4.1. The stopping time and excess. In this chapter $X_1, X_2, \cdots$ denote i.i.d. random variables with common distribution F and finite, positive mean μ, $0<\mu<\infty$; and $S_n = X_1+\cdots+X_n$, $n \geqq 1$, denote the partial sums. In addition, ξ_n, $n \geqq 1$, denote random variables for which $(X_1, \xi_1), \cdots, (X_n, \xi_n)$ are independent of X_k, $k>n$, for every $n \geqq 1$. The object is to extend aspects of renewal theory to

$$Z_n = S_n + \xi_n, \qquad n \geqq 1,$$

under smoothness conditions on ξ_n, $n \geqq 1$. It is convenient to let $Z_n = 0$, $\mathcal{A}_0 = \{\emptyset, \mathcal{X}\}$, where $\mathcal{X}$ is the sample space, and $\mathcal{A}_n = \sigma\{(X_k, \xi_k)\colon k \leqq n\}$, $n \geqq 1$. Thus, X_n, $n \geqq 1$, are independently adapted to $\mathcal{A}_n$, $n \geqq 1$. Next, let

$$\tau_a = \inf\{n \geqq 1\colon S_n > a\},$$

$$t = t_a = \inf\{n \geqq 1\colon Z_n > a\},$$

and

$$R_a = Z_{t_a} - a, \qquad a \geqq 0.$$

It is shown below that $t_a < \infty$ w.p. 1 for all $a \geqq 0$, so R_a, $a \geqq 0$, are well defined. These notations and assumptions are used throughout the chapter. We occasionally write t for t_a to avoid second order subscripts.

The process ξ_n, $n \geqq 1$, is said to be *slowly changing* if and only if:

(4.1) $$\frac{1}{n}\max\{|\xi_1|, |\xi_2|, \cdots, |\xi_n|\} \to 0$$

in probability as $n \to \infty$ and ξ_n, $n \geqq 1$, are uniformly continuous in probability—that is, for every $\varepsilon > 0$ there is a $\delta > 0$ for which

(4.2) $$P\left\{\max_{0 \leqq k \leqq n\delta} |\xi_{n+k} - \xi_n| > \varepsilon\right\} < \varepsilon \quad \text{for all } n \geqq 1.$$

See § 1.3. Observe that (4.1) holds if $\xi_n/n \to 0$ w.p. 1 as $n \to \infty$ and that (4.2) holds if ξ_n converges to a finite limit w.p. 1 as $n \to \infty$.

If ξ'_n, $n \geqq 1$, and ξ''_n, $n \geqq 1$, are two slowly changing sequences, then $\xi_n = \xi'_n + \xi''_n$, $n \geqq 1$, defines another slowly changing sequence.

Example 4.1. Let $Y_1, Y_2, \cdots$ be i.i.d. with a finite mean ν and a finite, positive variance.

(i) The sequence $\xi_n = n(\bar{Y}_n - \nu)^2$, $n \geqq 1$, is slowly changing. For $\xi_n/n = (\bar{Y}_n - \nu)^2 \to 0$ w.p. 1 as $n \to \infty$ by the strong law of large numbers (S.L.L.N.), and uniform continuity in probability follows from Kolmogorov's inequality as in Example 1.8.

(ii) If g is a positive, twice continuously differentiable function on $(-\infty, \infty)$, then $Z_n = ng(\bar{Y}_n)$ may be written in the form $Z_n = S_n + \xi_n$, where

$$S_n = ng(\nu) + ng'(\nu)(\bar{Y}_n - \nu), \qquad \xi_n = \tfrac{1}{2}g''(\nu_n)n(\bar{Y}_n - \nu)^2,$$

and ν_n denotes an intermediate point between $\bar{Y}_n$ and ν, $n \geqq 1$. Clearly, S_n, $n \geqq 1$, is a random walk with $X_i = g(\nu) + g'(\nu)(Y_i - \nu)$, $i \geqq 1$, and $\mu = g(\nu) > 0$. Next, $\xi_n/n = \frac{1}{2}g''(\nu_n)(\bar{Y}_n - \nu)^2 \to 0$ w.p. 1 as $n \to \infty$. And, finally, $g''(\nu_n)$ is uniformly continuous in probability, since $g''(\nu_n) \to g''(\nu)$ w.p. 1 as $n \to \infty$; so the products $\xi_n = \frac{1}{2}g''(\nu_n)n(\bar{Y}_n - \nu)^2$, $n \geqq 1$, are uniformly continuous in probability too. See Lemma 1.4.

(iii) It is only necessary for g to be positive and twice continuously differentiable in a neighborhood of ν, say $(\nu - \varepsilon, \nu + \varepsilon)$ with $\varepsilon > 0$. Let A_n be the event that $|\bar{Y}_n - \nu| < \varepsilon$, let $S_n = ng(\nu) + ng'(\nu)(\bar{Y}_n - \nu)$, and let $\xi_n = Z_n - S_n$, $n \geqq 1$. Then $\xi_n I_{A_n}$, $n \geqq 1$, are slowly changing, as in (ii), and $\xi_n I_{A_n'} \to 0$ w.p. 1 as $n \to \infty$, since $I_{A_n'} \to 0$ w.p. 1.

In the next lemma let $N = N_a = [a/\mu]$, the greatest integer in a/μ, $a \geqq 0$.

LEMMA 4.1. *If (4.1) holds, then $t_a < \infty$ for all $a \geqq 0$ w.p. 1 and $t_a/N_a \to 1$ in probability as $a \to \infty$. If $\xi_n/n \to 0$ w.p. 1, then $t_a/N_a \to 1$ w.p. 1 as $a \to \infty$.*

Proof. It follows directly from (4.1) and the S.L.L.N. that $(1/n) \max(|Z_1 - \mu|, \cdots, |Z_n - n\mu|) \to 0$ in probability as $n \to \infty$. In particular, $Z_n/n \to \mu$ in probability as $n \to \infty$, so there is a subsequence n_k, $k \geqq 1$, for which $Z_{n_k}/n_k \to \mu$ w.p. 1 as $k \to \infty$. It follows easily that $\sup_n Z_n = \infty$ w.p. 1, and this implies $t_a < \infty$ for all $a \geqq 0$ w.p. 1. Next, $t_a \to \infty$ w.p. 1 as $a \to \infty$. For t_a increase to a limit $t_\infty \leqq \infty$ w.p. 1 as $a \to \infty$, and $P\{t_\infty \leqq n\} = \lim_{a\to\infty} P\{\max_{k \leqq n} Z_k > a\} = 0$ for all $n \geqq 1$. Thus, if $\xi_n/n \to 0$ w.p. 1 as $n \to \infty$, then $Z_n/n \to \mu$ w.p. 1 as $n \to \infty$, and

$$\mu \leftarrow \frac{Z_{t_a - 1}}{t_a} \leqq \frac{a}{t_a} < \frac{Z_{t_a}}{t_a} \to \mu$$

w.p. 1 as $a \to \infty$, so that $t_a/a \to 1/\mu$ w.p. 1. Next, suppose only that (4.1) holds and let $\varepsilon > 0$. Then $t_a \leqq (1 - \varepsilon)N_a$ implies $Z_k - k\mu > a - k\mu$ for some $k \leqq (1 - \varepsilon)N_a$; and $a - (1 - \varepsilon)N_a\mu \geqq \varepsilon\mu N_a$, so that

$$P\{t_a \leqq (1 - \varepsilon)N_a\} \leqq P\left\{\max_{k \leqq N_a} |Z_k - k\mu| > \varepsilon\mu N_a\right\} \to 0$$

as $a \to \infty$. A similar calculation shows that $P\{t_a > (1 + \varepsilon)N_a\} \to 0$ to complete the proof.

LEMMA 4.2. *Suppose that F has a finite variance σ^2, that ξ_n, $n \geqq 1$, are slowly changing, and that $\xi_n/\sqrt{n} \to 0$ in probability as $n \to \infty$. Then*

$$t_a^* = \frac{t_a - N_a}{\sqrt{N_a}}$$

is asymptotically normal with mean 0 and variance $\mu^{-2}\sigma^2$ as $a \to \infty$.

Proof. By the definition of $t = t_a$,

$$\frac{1}{\sqrt{N_a}}(S_t - \mu t) = \frac{1}{\sqrt{N_a}}(a - \mu t) + \frac{1}{\sqrt{N_a}}(R_a - \xi_t), \qquad a \geqq 0. \tag{4.3}$$

The left side of (4.3) is asymptotically normal with mean 0 and variance σ^2, by Lemma 4.1 and Anscombe's theorem, so it suffices to show that the second term on the right tends to zero in probability. Now, $\xi_t/\sqrt{N_a} \to 0$ in probability by Anscombe's theorem and Lemma 4.1, since $\xi_n/\sqrt{n} \to 0$ in probability and $\xi_n/\sqrt{n}$ are u.c.i.p. Next,

$$0 \leqq R_a \leqq Z_t - Z_{t-1} = X_t + (\xi_t - \xi_{t-1}).$$

Since $E(X_1^2) < \infty$, $|X_t|/\sqrt{N_a} = \sqrt{t/N_a}\sqrt{X_t^2/t} \to 0$ in probability and $(\xi_t - \xi_{t-1})/\sqrt{N_a} \to 0$ in probability, as above.

THEOREM 4.1. *Suppose that F is nonarithmetic and that ξ_n, $n \geqq 1$, are slowly changing. Then R_a has limiting distribution H, as $a \to \infty$, where*

$$H\{dr\} = \frac{1}{E(S_\tau)} P\{S_\tau > r\}\, dr, \qquad r > 0,$$

and τ denotes the first ladder epoch of S_n, $n \geqq 1$. That is, R_a has the same limiting distribution as $S_{\tau_a} - a$.

Proof. Given $\varepsilon > 0$, let δ be as in (4.2), and let

$$N' = N'_a = \left[\frac{(1-\delta/4)a}{\mu}\right] \quad \text{and} \quad N'' = \left[\frac{(1+\delta/4)a}{\mu}\right];$$

observe that $(1+\delta)N' > N''$ for all large a. Next, let

$$B_a = \{t_{a-\sqrt{a}} > N'\} = \{\max(Z_1, \cdots, Z_{N'}) \leqq a - \sqrt{a}\}, \qquad a > 0.$$

Then $P(B_a) \to 1 \leftarrow P\{t_a \leqq N''\}$ as $a \to \infty$ by Lemma 4.1, since $a - \sqrt{a} \sim a$, $N'/N_a \to (1-\delta/4) < 1$, and $N''/N_a \to (1+\delta/4) > 1$. Observe that $a - Z_{N'} > \sqrt{a}$ on B_a, $a > 0$. Now let

$$S_n^\# = S_n - S_{N'}, \qquad n \geqq N'_a,$$

$$\tau_a^\# = \inf\{n > N' : S_n^\# > (a + \varepsilon - Z_{N'})\},$$

$$R_a^\# = S_{\tau_a^\#}^\# - (a + \varepsilon - Z_{N'}), \qquad a \geqq 0,$$

and let H_a be the distribution of $S_{\tau_a} - a$, $a \geqq 0$. Then the conditional distribution of $R_a^\#$, given $\mathscr{A}_{N'}$ is H_b, where $b = a + \varepsilon - Z_{N'}$, since the conditional distribution of $S_{N'+n}^\#$, $n \geqq 0$, is the same as the unconditional distribution of S_n, $n \geqq 0$. Moreover, since $b > \sqrt{a}$ on B_a, there is an $a_0 > 0$ for which $|H_b(r) - H(r)| \leqq \varepsilon$ for all $r > 0$ on B_a for $a > a_0$. Now let

$$C_a = B_a \cap \left\{t_a \leqq N'', \max_{N' \leqq k \leqq N''} |\xi_k - \xi_{N'}| \leqq \varepsilon\right\}, \qquad a \geqq 0.$$

Then $P(C_a)>1-\varepsilon$ for all large a by (4.2), since $P\{B_a, t_a \leqq N''\}\to 1$ as $a\to\infty$. It is shown below that, if $r>2\varepsilon$, then

$$C_a \cap \{R_a > r\} \subset \{\tau_a^{\#} = t_a, R_a^{\#} > r-2\varepsilon\}. \tag{4.4}$$

So

$$P\{R_a>r\}\leqq P\{R_a>r, C_a\}+P(C_a')\leqq P\{R_a^{\#}>r-2\varepsilon, B_a\}+\varepsilon$$
$$=\int_{B_a}[1-H_b(r-2\varepsilon)]\,dP+\varepsilon\leqq 1-H(r-2\varepsilon)+2\varepsilon$$

for all $r>2\varepsilon$ for sufficiently large a. Letting $a\to\infty$ and $\varepsilon\to 0$ in that order now shows that $\limsup P\{R_a>r\}\leqq 1-H(r)$ as $a\to\infty$ for all $r>0$.

A similar argument shows that $\liminf P\{R_a>r\}\geqq 1-H(r)$ as $a\to\infty$ for all $r>0$. The major difference is that ε is replaced by $-\varepsilon$ in the definition of $\tau_a^{\#}$, in which case $C_a\cap\{R_a^{\#}>r\}\subset\{t_a=\tau_a^{\#}, R_a>r-2\varepsilon\}$ for $r>2\varepsilon$. The details are omitted.

It remains to verify (4.4). If C_a occurs, then $S_k^{\#}=Z_k-Z_{N'}-(\xi_k-\xi_{N'})\leqq a-Z_{N'}+\varepsilon$ for $N'\leqq k<t_a$; so, $\tau_a^{\#}\geqq t_a$. If, in addition, $R_a>r>2\varepsilon$, then

$$S_{t_a}^{\#}=Z_{t_a}-Z_{N'}-(\xi_{t_a}-\xi_{N'})>a+2\varepsilon-Z_{N'}-\varepsilon=a-Z_{N'}+\varepsilon,$$

so that $\tau_a^{\#}=t_a$. Finally, if C_a occurs and $\tau_a^{\#}=t_a$, then $|R_a-R_a^{\#}|\leqq|\xi_{t_a}-\xi_{N'}|+\varepsilon\leqq 2\varepsilon$. These relations are illustrated in Fig. 4.1.

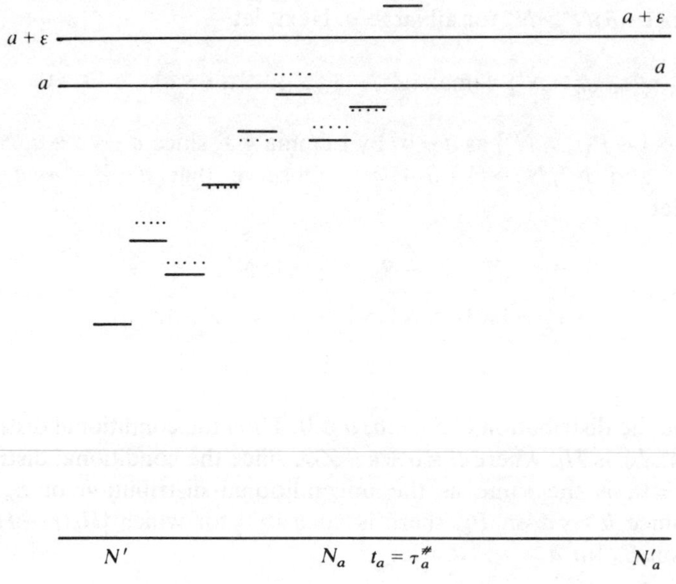

FIG. 4.1. *Graphs of Z_n and $S_n^{\#}+Z_{N'}$.* ——— Z_n; $\cdots\cdots$ $S_n^{\#}+Z_{N'}$.

Two simple modifications of the proof of Theorem 4.1 yield additional information. First, consider the joint distribution of R_a and t_a.

THEOREM 4.2. *Suppose that ξ_n, $n \geqq 1$, are slowly changing and that $\xi_n/\sqrt{n} \to 0$ in probability as $n \to \infty$. Suppose also that F is nonarithmetic and that F has a finite, positive variance σ^2. Then t_a^* and R_a are asymptotically independent as $n \to \infty$: that is,*

$$\lim_{a\to\infty} P\{t_a^* \leqq x, R_a \leqq r\} = \Phi\left(\frac{\mu x}{\sigma}\right) H(r), \qquad -\infty < x < \infty, \quad r \geqq 0,$$

where Φ denotes the standard normal distribution and H is as in Theorem 4.1.

Proof. Of course, it suffices to show that $P\{t_a^* > x, R_a > r\} \to [1 - \Phi(\mu x/\sigma)] \times [1 - H(r)]$ as $a \to \infty$ for $-\infty < x < \infty$ and $r > 0$. The proof is similar to that of Theorem 1 and is only sketched. Given x, r, and $\varepsilon > 0$, let δ, N', N'', B_a and C_a be as in the proof of Theorem 1, and lèt

$$m = m(a, x) = [Na + x\sqrt{Na}].$$

Then

$$\begin{aligned} P\{t_a^* > x, R_a > r\} &= P\{t_a > m, R_a > r\} \\ &\leqq \int_{t_a > m} P\{R_a > r, C_a | \mathcal{A}_m\}\, dP + P(C_a'), \qquad a > 0. \end{aligned}$$

By considering the process $S_k^{\#} = S_k - S_m$, $k \geqq 1$, one shows that $P\{R_a > r, C_a | \mathcal{A}_m\} \leqq 1 - H_b(r - 2\varepsilon)$ for $r > 2\varepsilon$, where $b = a + \varepsilon - Z_m$. Letting $a \to \infty$ then shows that

$$\limsup_{a\to\infty} P\{t_a^* > x, R_a > r\} \leqq [1 - H(r - 2\varepsilon)]\left[1 - \Phi\left(\frac{\mu x}{\sigma}\right)\right] + \varepsilon,$$

since t_a^* is asymptotically normal with mean 0 and variance $\mu^{-2}\sigma^2$ and $P(C_a') \leqq \varepsilon$ for all large a. A similar lower bound may be obtained to complete the proof.

Next, consider the arithmetic case. It is convenient to write $[x]$ and $\langle x \rangle = x - [x]$ for the integer and fractional parts of x, $-\infty < x < \infty$. Observe that if ξ is a random variable with distribution function G, then the distribution of $\langle \xi \rangle$ is

$$G_0\{A\} = \sum_{j=-\infty}^{\infty} G\{j + A\}, \tag{4.5}$$

for Borel sets $A \subset [0, 1)$.

THEOREM 4.3. *Suppose that F is arithmetic with span $d = 1$. Suppose also that ξ_n, $n \geqq 1$, are slowly changing and that ξ_n converges in distribution to a random variable ξ having a continuous distribution G. Then $R_a = Z_t - a$ has a limiting distribution K as $a \to \infty$ through integers. The distribution K is determined by the conditions*

$$K\{j + A\} = H\{j + 1\} G_0\{A\}$$

for $j \geqq 0$ and Borel sets $A \subset [0, 1)$, where G_0 denotes the distribution of $\langle \xi \rangle$ and

H denotes the asymptotic distribution of $S_{\tau_a} - a$; that is, $H\{k\} = (1/E(S_\tau))P\{S_\tau \geqq k\}$, $k \geqq 1$, where τ denotes the first ladder height.

The conclusion to Theorem 4.2 may be written in the alternative form: R_a converges in distribution to $R_0 + \langle \xi_0 \rangle - 1$, where R_0 and ξ_0 are independent, R_0 has distribution H, and ξ_0 has distribution G. It is interesting that the asymptotic distribution of R_a in the nonlinear case ($\xi_n \Rightarrow \xi \neq 0$) is different from that of the linear case.

Proof. Given ε, $0 < \varepsilon < \frac{1}{4}$, construct N', N'', B_a and $S_n^\# = S_n - S_{N'}$, $n \geqq N'$, as in Theorem 4.1, but include $\varepsilon < \langle \xi_{N'} \rangle < 1 - \varepsilon$ in C_a and let

$$\tau_a^\# = \inf\{n \geqq N' : S_n^\# > (a - 1 - [Z_{N'}])\}$$

and

$$R_a^\# = S_{\tau_a^\#}^\# - (a - 1 - [Z_{N'}]), \qquad a > 0. \tag{4.6}$$

Then the conditional distribution of $R_a^\#$, given $\mathscr{A}_{N'}$, is H_b, where $b = a - 1 - [Z_{N'}]$. As in the proof of Theorem 1, one shows that $\tau_a^\# = t_a$ with high probability. Then

$$R_a = R_a^\# + (\langle \xi_{N'} \rangle - 1) + (\xi_t - \xi_{N'}).$$

The last term on the right is small with high probability, and the first two have the limiting distribution described in Theorem 4.3. The reader may supply the details.

Remarks and references. Theorems 4.1 and 4.2 are taken from Lai and Siegmund (1977). Theorem 4.3 is taken from Lalley (1980). Lalley also determines the limiting joint distribution of t_a^* and R_a in the discrete case. Hagwood (1980) determines the limiting distribution of R_a in the discrete case, using local methods to be described in § 5.2.

Decompositions of the form $Z_n = S_n + \xi_n$, $n \geqq 1$, were used by Pollak and Siegmund (1975) in a related context.

4.2. The expected stopping time. In this section approximations to the mean and variance of t_a are developed. The notation and standing assumptions of § 4.1 are used throughout. In particular, $N_a = [a/\mu]$, $a > 0$.

THEOREM 4.4. *Suppose that F has a finite variance σ^2, that ξ_n, $n \geqq 1$, satisfy condition* (4.1), *and that*

$$\sum_{n=1}^{\infty} P\{\xi_n \leqq -n\varepsilon\} < \infty \quad \textit{for some } \varepsilon, \quad 0 < \varepsilon < \mu. \tag{4.7}$$

Then

$$\frac{E(t_a)}{N_a} \to 1 \quad \textit{as } a \to \infty.$$

Proof. Let ε be as in (4.7), $0 < \varepsilon < \mu$, let $\delta > 0$ be so small that $\varepsilon + \delta < \mu$ and let $K_a = [a/(\mu - \varepsilon - \delta)] + 1$. Then, for $n > K_a$, $a - n\mu < -n(\varepsilon + \delta)$, so that

$$P\{t > n\} \leqq P\{S_n - n\mu + \xi_n \leqq a - n\mu\} \leqq P\{S_n - n\mu \leqq -n\delta\} + P\{\xi_n \leqq -n\varepsilon\}. \tag{4.8}$$

Let γ_n denote the right side of (4.8), $n \geqq 1$. Then γ_n, $n \geqq 1$, are independent of $a \geqq 0$, and γ_n, $n \geqq 1$, are summable, by Corollary 2.5 and (4.7). Now,

$$(4.9)\qquad \begin{aligned} \int_{t_a > 2K_a} t_a \, dP &\leqq 2 \int_{t_a > 2K_a} (t_a - K_a) \, dP \leqq 2 \int_{t_a > K_a} (t_a - K_a) \, dP \\ &= 2 \sum_{n=K_a}^{\infty} \gamma_n = o(1) \end{aligned}$$

as $a \to \infty$. It follows that t_a/a, $a \geqq 0$, are uniformly integrable and, since $t_a/a \to 1/\mu$ in probability as $a \to \infty$, $E(t_a/a) \to 1/\mu$.

COROLLARY 4.1. *Suppose that F has a finite variance σ^2 and that ξ_n, $n \geqq 1$, satisfy conditions* (4.1) *and* (4.7). *If*

$$E[(R_a - \xi_{t_a})^2] = o(N_a) \quad \textit{as } a \to \infty,$$

then

$$E[(t_a - N_a)^2] \sim \mu^{-2}\sigma^2 N_a \quad \textit{as } a \to \infty.$$

Proof. By definition of t_a and R_a,

$$\left(t_a - \frac{a}{\mu}\right) = \frac{-1}{\mu}(S_{t_a} - \mu t_a) + \frac{1}{\mu}(R_a - \xi_{t_a}), \qquad a > 0.$$

Now,

$$E[(S_{t_a} - \mu t_a)^2] = \sigma^2 E(t_a) \sim \sigma^2 N_a \quad \text{as } a \to \infty,$$

by Wald's lemma and Theorem 4.4, and $E[(R_a - \xi_{t_a})^2] = o(N_a)$ as $a \to \infty$, by assumption. The corollary follows easily.

The main result of this section gives a more detailed approximation to $E(t_a)$, under additional conditions. In it, suppose that there are $\mathcal{A}_n$-measurable events A_n, $n \geqq 1$, constants l_n, $n \geqq 1$, and $\mathcal{A}_n$-measurable random variables V_n, $n \geqq 1$, for which:

$$(4.10)\qquad \sum_{n=1}^{\infty} P\left(\bigcup_{k=n}^{\infty} A_k'\right) < \infty,$$

$$(4.11)\qquad \xi_n = l_n + V_n \quad \text{on } A_n, \quad n \geqq 1,$$

$$(4.12)\qquad \sup_{n \geqq 1} \max_{0 \leqq k \leqq n\delta} |l_{n+k} - l_n| \to 0 \quad \text{as } \delta \to 0,$$

$$(4.13)\qquad \max_{0 \leqq k \leqq n} |V_{n+k}|, \ n \geqq 1, \text{ are uniformly integrable},$$

$$(4.14)\qquad \sum_{n=1}^{\infty} P\{V_n \leqq -n\varepsilon\} < \infty \quad \text{for some } \varepsilon, \quad 0 < \varepsilon < \mu,$$

$$(4.15)\qquad V_n \text{ converges in distribution to a random variable } V,$$

and

$$P\{t_a \leqq \varepsilon N_a\} = o\left(\frac{1}{N_a}\right) \quad \text{as } a \to \infty, \quad \varepsilon > 0, \tag{4.16}$$

where $N_a = [a/\mu]$, $a \geqq 0$. Observe that if (4.10)–(4.12) hold and if V_n, $n \geqq 1$, are slowly changing, then ξ_n, $n \geqq 1$, are also slowly changing.

THEOREM 4.5. *Suppose that F has finite, positive variance σ^2. Suppose also that conditions* (4.10)–(4.16) *are satisfied and that V_n, $n \geqq 1$, are slowly changing. If F is nonarithmetic, then*

$$E(t_a) = \frac{1}{\mu}(a + \rho - l_{N_a} - E(V)) + o(1) \quad \text{as } a \to \infty, \tag{4.17}$$

where $\rho = E(S_\tau^2)/2E(S_\tau)$ denotes the mean of the asymptotic distribution H of R_a. If F is arithmetic with span $d > 0$, if $l_n = 0$, $n \geqq 1$ and if V has a continuous distribution, then (4.17) *holds as $a \to \infty$ through multiples of d with $\rho = E[S_\tau(S_\tau + d)]/2E(S_\tau)$.*

The proof of Theorem 4.5 is presented below.

In applications, condition (4.16) is often the most troublesome. While complicated, the remaining conditions are not especially restrictive. For example, when $A_n = \mathscr{X}$, the sample space, and $l_n = 0$, $n \geqq 1$, then they require that $\xi_n = V_n$ have a limiting distribution, that $\max_{0 \leqq k \leqq n} |\xi_{n+k}|$ be uniformly integrable, and that ξ_n satisfy the mild boundedness condition (4.7). Including the events A_n, $n \geqq 1$, in the formulation avoids requiring that ξ_n be integrable, and including the constants l_n, $n \geqq 1$, avoids requiring that ξ_n be bounded. Observe that (12) holds of $l_n = b(\log n)^\alpha + c_n$, where $0 < \alpha \leqq 1$ and c_n converges to a finite limit as $n \to \infty$. In fact, (4.12) requires $l_n = O(\log n)$.

Example 4.2. If F has a finite variance σ^2 and if $\xi_n = 0$ for all $n \geqq 1$, then (4.16) holds for any ε, $0 < \varepsilon < 1$. To see why, let $0 < \varepsilon < 1$, let $K = K_a = [(1-\varepsilon)N_a]$ and let $M = \max(S_1 - \mu, S_2 - 2\mu, \cdots, S_K - K\mu)$. Then

$$\begin{aligned} P\{t_a \leqq K\} &= P\left\{\max_{0 \leqq k \leqq K} S_k > a\right\} \leqq P\{M > \varepsilon\mu K\} \\ &\leqq \left(\frac{1}{K\varepsilon\mu}\right)^2 \int_{M > \varepsilon\mu K} (S_K - K\mu)^2 \, dP \end{aligned} \tag{4.18}$$

by the martingale inequality. Using $D(S_K^2) = K\sigma^2$, one sees that $P\{M > \varepsilon\mu K\} \to 0$ as $a \to \infty$; then, using the uniform integrability of $S_K^* = (S_K - K\mu)/\sigma\sqrt{K}$, one sees that the last integral in (4.18) is $o(1/K)$ as $a \to \infty$. That is, $P\{t_a \leqq K\} = o(1/K)$ as $a \to \infty$.

Example 4.3. Let $Y_1, Y_2, \cdots$ be i.i.d. with mean $\nu > 0$ and finite αth moment $E|Y_1|^\alpha$ for some $\alpha > 2$, and let $Z_n = \frac{1}{2}n\bar{Y}_n^2$, $n \geqq 1$. Then conditions (4.10)–(4.16) are satisfied, with $A_n = \mathscr{X}$ and $l_n = 0$, $n \geqq 1$. To see why, write $Z_n = S_n + \xi_n$ with $S_n = \frac{1}{2}n\nu^2 + n\nu(\bar{Y}_n - \nu)$ and $\xi_n = \frac{1}{2}n(\bar{Y}_n - \nu)^2$, $n \geqq 1$. Then (4.10)–(4.12) are satisfied with $A_n = \mathscr{X}$ and $V_n = \xi_n$, $n \geqq 1$; (4.14) holds trivially, since $\xi_n \geqq 0$, $n \geqq 1$; and the asymptotic distribution of $V_n = \xi_n$ is a multiple of chi-squared in (4.15).

For (4.13) one has

$$P\Big\{\max_{0\leq k\leq n} \xi_{n+k} > y\Big\} \leq P\Big\{\max_{0\leq k\leq 2n} k|\bar{Y}_k - \nu| > \sqrt{2ny}\Big\}$$
$$\leq \left(\frac{1}{2ny}\right)^{\alpha/2} E|2n(\bar{Y}_{2n}-\nu)|^{\alpha} \leq Cy^{-\alpha/2}$$

for all $y>0$ and $n\geq 1$ for some constant C by the martingale inequality and von Bahr's theorem. For (4.16) observe first that $\sqrt{2an-n\nu}>\sqrt{an}$ for all $n\leq \varepsilon N_a$ for sufficiently small $\varepsilon>0$. For such ε, $t_a\leq \varepsilon N_a$ implies that $n|\bar{Y}_n-\nu|>\sqrt{an}$ for some $n\leq \varepsilon N_a$, and this implies that

$$B_k = \Big\{\max_{0\leq n\leq 2^k} n|\bar{Y}_n - \nu| > \sqrt{a2^{k-1}}\Big\}$$

occurs for some $k\leq K_a = [\log_2(\varepsilon N_a)]+1$. As above, one finds $P(B_k)\leq Ca^{-\alpha/2}$ for all $a>0$ for some C. So $P\{t_a\leq \varepsilon N_a\}\leq CK_a a^{-\alpha/2}=o(1/a)$ as $a\to\infty$.

Proof of Theorem 4.5. Observe that (4.12) requires $l_n/n\to 0$ as $n\to\infty$. It follows that (4.14) implies (4.7), and (4.7) implies (4.9). Combining (4.9) and (4.16) then shows the existence of ε_1, ε_2 for which $0<\varepsilon_1<1<\varepsilon_2<\infty$ for which

$$N_a P\{t_a\leq \varepsilon_1 N_a\} + \int_{t_a\geq \varepsilon_2 N_a} t_a\, dP \to 0, \qquad a\to\infty. \tag{4.19}$$

Let

$$N' = [\varepsilon_1 N_a],\quad N'' = [\varepsilon_2 N_a]+1,\quad b = a - l_{N_a}$$

and

$$B_a = \{N'\leq t_a\leq N'',\ \tau_b > N'\}\cap \bigcap_{k=N'}^{\infty} A_k,$$

where τ_a, $a>0$ are the first passage times for the random walk S_n, $n\geq 0$. Then $P(B_a')=o(1/N_a)$ as $a\to\infty$, by (4.10), (4.19) and Example 4.3, since $b\sim a$. It follows that

$$\int_{B_a} S_{t_a}\, dP = \int_{B_a} (a+R_a-\xi_{t_a})\, dP = a + \int_{B_a} (R_a - \xi_{t_a})\, dP + o(1), \qquad a\to\infty.$$

The theorem then follows by substituting for the integrals the approximations developed in Lemmas 4.3, 4.4 and 4.5 below.

LEMMA 4.3.

$$\int_{B_a} \xi_{t_a}\, dP = l_{N_a} + E(V) + o(1) \quad \textit{as } a\to\infty.$$

Proof. It follows from Lemma 4.1 and Anscombe's theorem that

$$(\xi_{t_a} - l_{N_a})I_{B_a} = (\xi_{t_a} - \xi_{N_a})I_{B_a} + V_{N_a} I_{B_a}$$

converges in distribution to V. Moreover $(\xi_{t_a} - l_{N_a})I_{B_a}$ is dominated by

$$M_a = \max_{N' \leqq k \leqq N''} |\xi_k - l_{N_a}| I_{B_a},$$

which is uniformly integrable by (4.12) and (4.13).

LEMMA 4.4.

$$\int_{B_a} R_a \, dP = \rho + o(1) \quad \text{as } a \to \infty,$$

if either F is nonarithmetic or F is arithmetic with span d and $a \to \infty$ *through multiples of d.*

Proof. Recall that $b = a - l_{N_a}$ and observe that

$$R_a = (S_{t_a} - b) + (\xi_{t_a} - l_{N_a}).$$

Since the second term on the right was shown to be uniformly integrable over B_a in the proof of Lemma 4.3, it suffices to show that the first is. Let

$$C_a = C_a(r) = \{M_a \leqq r\}, \qquad a, r > 0.$$

Then

$$P\{B_a, S_{t_a} - b > 2r\} \leqq P\{B_a \cap C_a, S_{t_a} - b > 2r\} + P(C'_a) \tag{4.20}$$

and

$$\begin{aligned} P\{B_a \cap C_a, S_{t_a} - b > 2r\} &= \sum_{n=N'}^{N''} P\{B_a \cap C_a, t_a = n, S_n > b + 2r\} \\ &\leqq \sum_{n=N'}^{N''} P\{B_a \cap C_a, t_a \geqq n, S_n > b + 2r\}. \end{aligned}$$

Now, for $N' \leqq n \leqq N''$, B_a, C_a, and $t_a \geqq n$ imply that $\tau_{b+r} \geqq n$, for they imply that $\tau_{b+r} \geqq \tau_b > N'_a$ and that

$$S_k = Z_k - \xi_k \leqq a - \xi_k = b - (\xi_k - l_{N_a}) \leqq b + r, \qquad N'_a \leqq k \leqq n.$$

So

$$\begin{aligned} P\{B_a \cap C_a, S_{t_a} - b > 2r\} &\leqq \sum_{n=N'_a}^{N''_a} P\{\tau_{b+r} \geqq n, S_n > b + 2r\} \\ &\leqq P\{S_{\tau_{b+r}} - (b+r) > r\} \leqq \sup_{c>0} P\{S_{\tau_c} - c > r\}, \end{aligned}$$

which is integrable over $r > 0$ by Theorem 2.4. The uniform integrability of $R_a I_{B_a}$ now follows from (4.19), (4.20), and the uniform integrability of M_a, $a > 0$.

LEMMA 4.5.

$$\int_{B_a} S_{t_a} \, dP = \mu E(t_a) + o(1) \quad \text{as } a \to \infty.$$

Proof. By Wald's lemma

$$\int_{B_a} S_{t_a}\,dP = \mu E(t_a) - \int_{B'_a} S_{t_a}\,dP, \qquad a>0.$$

Now,

$$\int_{B'_a} S_{t_a}\,dP = \int_{B'_a} \mu t_a\,dP + \int_{B'_a} (S_{t_a} - \mu t_a)\,dP, \qquad a>0.$$

By (4.19)

$$\int_{B'_a} \mu t_a\,dP \to 0 \quad \text{as } a\to\infty,$$

and

$$\left|\int_{B'_a} (S_{t_a} - \mu t_a)\,dP\right| \leqq \sqrt{P(B'_a)E[(S_{t_a} - \mu t_a)^2]}$$
$$= \sqrt{o(1/N_a)\sigma^2 E(t_a)} = o(1) \quad \text{as } a\to\infty.$$

Remarks and references. Theorem 4.5 is adapted from Hagwood and Woodroofe (1981).

Lai and Siegmund (1979) develop another approach to the uniform integrability of R_a and asymptotic expansions for $E(t_a)$. They form the renewal measure

$$V\{J\} = \sum_{n=1}^{\infty} P\{Z_n \in J\},$$

and show

$$\lim_{a\to\infty} V\{a+J\} = \frac{m_0\{J\}}{\mu}$$

for finite intervals J, under the following conditions: F is nonarithmetic and for some α, $\frac{1}{2} < \alpha \leqq 1$, and all $\varepsilon > 0$, $E|X_1|^{2/\alpha} < \infty$,

$$\sum_{n=1}^{\infty} P\{|\xi_n| > n^\alpha \varepsilon\} < \infty,$$

$$\sum_{n<j<n+\delta n^\alpha} P\{|\xi_j - \xi_n| > \varepsilon\} < \varepsilon \quad \text{for all } n \geqq 1,$$

for all sufficiently small $\delta > 0$, depending on ε. Once this nonlinear renewal theorem has been established, proofs of uniform integrability for R_a and expansions for $E(t_a)$ may be constructed along the lines of Theorem 2.4 and Corollary 2.2. The resulting conditions appear to be slightly more complicated than (4.10)–(4.16) and to require slightly more in the way of moment conditions, but the conditions imposed on the sequence of constants is weaker than (4.12).

CHAPTER 5

Local Limit Theorems

5.1. Conditional probabilities. In this chapter the nature of the convergence in Theorems 4.1–4.3 is examined in more detail in the special case (5.1) below. Throughout, $Y_1, Y_2, \cdots$ denote i.i.d. random variables with common distribution G, finite mean ν, and finite positive variance τ^2, $\mathcal{Y}$ denotes an interval for which $G\{\mathcal{Y}\}=1$ and $\nu \in \mathcal{Y}^0$, the interior of $\mathcal{Y}$, and

$$(5.1) \qquad Z_n = n\Delta(\bar{Y}_n), \qquad n \geqq 1,$$

where Δ is a continuous function on $\mathcal{Y}$ with $\Delta(\nu)>0$. Recall from Example 4.1 that if Δ is twice continuously differentiable near ν, then $Z_n = S_n + \xi_n$, $n \geqq 1$, where ξ_n, $n \geqq 1$, are slowly changing and

$$(5.2) \qquad S_n = n\Delta(\nu) + n\Delta'(\nu)(\bar{Y}_n - \nu), \qquad n \geqq 1,$$

is a random walk, and observe that the mean and variance of $X_1 = S_1$ are $\mu = \Delta(\nu) > 0$ and $\sigma^2 = \Delta'(\nu)^2\tau^2$.

Let m be a fixed positive integer and let

$$t_a = \inf\{n \geqq m: Z_n > a\}, \qquad a \geqq 0.$$

The main result of this section gives an approximation to the conditional probability that $t_a \geqq n$, given that $Z_n \approx a$. That is, an approximation to

$$
\begin{aligned}
u_a(n, y) &= P\{t_a \geqq n \mid \bar{Y}_n = y\} \\
(5.3) \qquad &= P\{Z_k \leqq a,\ m \leqq k \leqq n-1 \mid \bar{Y}_n = y\} \\
&= P\{Z_n - Z_{n-k} \geqq n\Delta(y) - a,\ 1 \leqq k \leqq n-m \mid \bar{Y}_n = y\}
\end{aligned}
$$

is developed under the assumption that n and a are large and $n\Delta(y) \approx a$. As (5.3) indicates, the approach is to compute the conditional probability of crossing the boundary a, looking backward along the sequence $Z_{n-1}, \cdots, Z_m$.

The approximation requires some conditions on G. Briefly, the conditions allow the use of the local central limit theorem. Recall that a random variable Y has a lattice distribution if and only if $Y-c$ has an arithmetic distribution for some c, $-\infty < c < \infty$, in which case the lattice span of Y is the maximum of the arithmetic spans of $Y-c$ for such c. For example, if $Y = \pm 1$ w.p. $\frac{1}{2}$ each, then Y is arithmetic with arithmetic span 1; but the lattice span of Y is 2, since $Y+1=0$ or 2 w.p. $\frac{1}{2}$ each.

Condition S (*smoothness*). Some power of the characteristic function of Y_1 is integrable with respect to Lebesgue measure on $(-\infty, \infty)$.

Condition A (*arithmetic*). Y_1 is integer-valued with lattice span 1.

If condition S is satisfied, then the sum $n\bar{Y}_n = Y_1 + \cdots + Y_n$ has a bounded continuous density g_n with respect to Lebesgue measure for sufficiently large n, say $n \geqq n_0$, by Fourier inversion. Thus

$$g_n(s) = \frac{d}{ds} P\{Y_1 + \cdots + Y_n \leqq s\}, \qquad -\infty < s < \infty, \quad n \geqq n,$$

and

$$g_n(s) = \left(\frac{1}{\sqrt{n}}\right)\Phi'\left[\frac{s - n\nu}{\tau\sqrt{n}}\right] + o\left(\frac{1}{\sqrt{n}}\right) \tag{5.4}$$

uniformly in s, $-\infty < s < \infty$ as $n \to \infty$, where Φ denotes the standard normal distribution function. If condition A is satisfied, then (5.4) holds uniformly in integral s, $-\infty < s < \infty$, with

$$g_n(s) = P\{Y_1 + \cdots + Y_n = s\}, \qquad -\infty < s < \infty, \quad n \geqq 1.$$

See Feller (1966, pp. 488–491) for both assertions.

Recall that probability distributions F_n, $n \geqq 1$, defined on the Borel sets of $(-\infty, \infty)^k$, where $k \geqq 1$, converge weakly to a probability distribution F if and only if

$$F_n\{B\} \to F\{B\} \quad \text{as } n \to \infty, \tag{5.5}$$

for all Borel sets B for which $F\{\partial B\} = 0$, where ∂B denotes the topological boundary of B. By contrast, F_n is said to *converge strongly* to F if and only if (5.5) holds uniformly in all Borel sets B. A sufficient condition for the strong convergence of F_n to F is that F_n, $n \geqq 1$, and F have densities f_n, $n \geqq 1$, and f with respect to a dominating measure λ and that

$$\int |f_n - f| \, d\lambda \to 0 \quad \text{as } n \to \infty, \tag{5.6}$$

for $|F_n\{B\} - F\{B\}| \leqq \int_B |f_n - f| \, d\lambda \leqq \int |f_n - f| \, d\lambda$ for all B. In fact (5.6) is both necessary and sufficient for the strong convergence of F_n to F. In turn, (5.6) may be deduced from the pointwise convergence of f_n to f and Scheffe's theorem (1947): Let λ be a measure and let f_n, $n \geqq 1$, and f be nonnegative measurable functions. If $f_n \to f$ almost everywhere (λ) and $\int f_n \, d\lambda \to \int f \, d\lambda$ as $n \to \infty$, then (5.6) holds. See Lehmann (1959, pp. 351–352) for a simple proof.

For example, if condition S is satisfied, then the distributions of $W_n = \sqrt{(n/\tau)}(\bar{Y}_n - \nu)$ converge strongly to the standard normal density.

Lemma 5.1 details the relevant properties of the conditional distributions of $Y_1, \cdots, Y_n$, given $\bar{Y}_n$.

LEMMA 5.1. *Suppose that either* S *or* A *is satisfied. Define* q_{nk} *and* Q_{nk} *by*

$$q_{nk}(y_1, \cdots, y_k | y) = \begin{cases} \dfrac{g_{n-k}[ny - \sum_{j=1}^k y_j]}{g_n(ny)}, & g_n(ny) > 0, \\ 1, & g_n(ny) = 0 \end{cases}$$

and

$$Q_{nk}\{B|y\} = \int \cdots \int_B q_{nk}(y_1, \cdots, y_k|y) G\{dy_1\} \times \cdots \times G\{dy_k\}$$

for Borel sets $B \subset (-\infty, \infty)^k$, $1 \leqq k \leqq n-1$, *and* $n \geqq n_0$. *Then*:

(i) Q_{nk} *is (a version of) the conditional distribution of* $Y_1, \cdots, Y_k$ *given* $\bar{Y}_n$ *for* $1 \leqq k \leqq n-1$, $n \geqq n_0$.

(iii) *If* $b > 0$ *and* $0 < \delta < 1$, *then there is a constant* $C = C(b, \delta)$ *for which* $q_{nk} \leqq C$ *for all (integral)* $y_1, \cdots, y_k$ *for all* $k \leqq (1-\delta)n$, *whenever* $|y - \nu| \leqq b/\sqrt{n}$ *and* $G_1 \times \cdots \times G$ *for fixed k.*

(iii) *If* $b > 0$ *and* $0 < \delta < 1$, *then there is a constant* $C = C(b, \delta)$ *for which* $q_{nk} \leqq C$ *for all (integral)* $y_1, \cdots, y_k$ *for all* $k \leqq (1-\delta)n$, *whenever* $|y - \nu| \leqq b/\sqrt{n}$ *and* $n \geqq n_0$.

Proof. That Q_{nk} is the conditional distribution of $Y_1, \cdots, Y_k$, given $\bar{Y}_n$, is clear. The convergence in (ii) follows easily from the local limit theorem, which implies that $g_{n-k}(ny - x)/g_n(ny) \to 1$ as $n \to \infty$ for fixed x and k. The inequality in (iii) also follows from the local limit theorem; for (5.4) implies that $\sqrt{n} g_n(ny)$ is bounded above for all y, $-\infty < y < \infty$, and bounded away from zero when $|y - \nu| \leqq b/\sqrt{n}$ (and ny is an integer, in the arithmetic case).

The conditional distributions Q_{nk} are used in all conditional probabilities which appear below.

From equation (5.3), one may write

$$u_a(n, y) = v_a[n, y, n\Delta(y) - a], \qquad n \geqq 1, \quad a \geqq 0,$$

where

$$v_a(n, y, r) = P\{Z_n - Z_{n-k} \geqq r, 1 \leqq k \leqq n - m | \bar{Y} = y\}, \qquad -\infty < r < \infty.$$

Let

$$v(r) = P\{S_k \geqq r, \text{ for all } k \geqq 1\}, \qquad -\infty < r < \infty,$$

where S_k, $k \geqq 1$, denotes the random walk (5.2). Then $v_a(n, y, r)$ and $v(r)$ are nonincreasing, left continuous functions of r for $-\infty < y < \infty$ and $n \geqq n_0$. It is easily seen that the continuity set $\mathscr{C}_v$ of v consists of all $r \geqq 0$ for which $P\{\min(S_1, \cdots, S_k) = r\} = 0$ for all $k \geqq 1$. In particular, v is continuous if condition S is satisfied.

THEOREM 5.1. *Suppose that* Δ *is positive and continuously differentiable near* ν *with* $\Delta'(\nu) \neq 0$. *Suppose also that either condition* S *or* A *holds. Let* $y = y_a \to \nu$ *and* $n = n_a \to \infty$ *as* $a \to \infty$ *with*

$$y - \nu = O\left(\frac{1}{\sqrt{n}}\right) \quad \text{and} \quad n\Delta(y) - a = O(1),$$

and suppose that ny *are integers in the arithmetic case. Then* $\lim v_a(n, y, r) = v(r)$ *for all* $r \in \mathscr{C}_v$.

COROLLARY 5.1. *If* $n\Delta(y) - a \to r \in \mathscr{C}_v$, *then*

$$u_a(n, y) \to v(r). \tag{5.7}$$

Proof. The corollary is a clear consequence of the theorem. To prove the theorem, let

$$Y_{nk} = Y_k - \bar{Y}_n, \qquad T_{nk} = Y_{n1} + \cdots + Y_{nk}, \qquad 1 \leqq k \leqq n.$$

Then the conditional distribution of $(Y_{n1}, \cdots, Y_{nk})$, given $\bar{Y}_n = y$, converges weakly to the unconditional distribution of $(Y_1 - \nu, \cdots, Y_k - \nu)$ for fixed k, by Lemma 5.1. Next, given $\bar{Y}_n = y$,

$$Z_n - Z_{n-k} = (n-k)[\Delta(\bar{Y}_n) - \Delta(\bar{Y}_{n-k})] + k\Delta(y)$$

and

$$(n-k)[\bar{Y}_n - \bar{Y}_{n-k}] = -\sum_{j=1}^{n-k} Y_{nj} = \sum_{j=n-k+1}^{n} Y_{nj}, \qquad 1 \leqq k \leqq n, \tag{5.8}$$

since $Y_{n1} + \cdots + Y_{nn} = 0$. Next, since $Y_{n1}, \cdots, Y_{nn}$ are conditionally exchangeable, given $\bar{Y}_n = y$, the last sum in (5.8) has the same conditional distribution as T_{nk}; in fact, $Z_n - Z_{n-1}, \cdots, Z_n - Z_1$ have the same joint conditional distribution as

$$W_{nj} = (n-j)\left[\Delta(y) - \Delta\left(y - \frac{1}{n-j} T_{nj}\right)\right] + j\Delta(y), \qquad 1 \leqq j \leqq n-1.$$

Moreover, for fixed k, the conditional distributions of $(W_{n1}, \cdots, W_{nk})$ converge weakly to the unconditional distribution of $(S_1, \cdots, S_k)$. For $W_{nj} = h_{nj}(T_{nj})$, where $h_{nj}(t) \to \Delta'(\nu)t + j\Delta(\nu)$ uniformly in t on compact subintervals of $(-\infty, \infty)$ for each j, and the conditional distributions of $(Y_{n1}, \cdots, Y_{nk})$ converge to the distribution of $(Y_1 - \nu, \cdots, Y_k - \nu)$ for fixed k, as above. See Billingsley (1968, pp. 30–31). So the conditional distributions of $(Z_n - Z_{n-1}, \cdots, Z_n - Z_{n-k})$ converge to the distribution of $(S_1, \cdots, S_k)$ for fixed k. It follows that

$$v_{a,k}(n, y, r) \underset{\text{def}}{=} P\{Z_n - Z_{n-j} \geqq r, 1 \leqq j \leqq k \mid \bar{Y}_n = y\}$$

$$\to P\{S_j \geqq r, 1 \leqq j \leqq k\} \underset{\text{def}}{=} v_k(r)$$

for all continuity points $r \in \mathscr{C}_v$ for each fixed k. Since $v_k(r) \to v(r)$ as $k \to \infty$ for all $r \geqq 0$, it now suffices to show that the difference $\varepsilon_{a,k} = v_{a,k}(n, y, r) - v_a(n, y, r)$ is small for large a, when k is sufficiently large.

To estimate $\varepsilon_{a,k}$ first observe that for any δ, $0 < \delta < \frac{1}{2}$,

$$\begin{aligned} 0 \leqq \varepsilon_{a,k} &\leqq P\{Z_n - Z_{n-j} < r, \exists j \in (k, n-m] \mid \bar{Y}_n = y\} \\ &\leqq P\{Z_n - Z_{n-j} < r, \exists j \in (k, n\delta] \mid \bar{Y}_n = y\} \\ &\quad + P\{Z_n - Z_{n-j} < r, \exists j \in (n\delta, n-m] \mid \bar{Y}_n = y\} \\ &= \varepsilon_{a,1}(k, \delta) + \varepsilon_{a,2}(\delta), \quad \text{say.} \end{aligned}$$

The second of these two terms is easily bounded. Indeed, letting $b = n\Delta(y) - r$ and replacing $n-j$ by j, one finds that

$$\varepsilon_{a,2}(\delta) = P\{Z_j > b, \exists j \in [m, n(1-\delta)) | \bar{Y}_n = y\} \tag{5.9}$$
$$\leqq CP\{Z_j > b, \exists j \in [m, n(1-\delta))\} = CP\{t_b \leqq n(1-\delta)\},$$

by Lemma 5.1 (iii), and the last term in (5.9) approaches zero as $a \to \infty$, since $a \sim b$, $t_b/b \to 1/\mu = 1/\Delta(\nu)$ in probability, and $n \sim a/\Delta(\nu)$ as $a \to \infty$.

Since Δ is differentiable at ν, there are N and $\gamma > 0$ for which

$$\left|(n-j)\left[\Delta(y) - \Delta\left(\frac{1}{n-j} T_{nj}\right)\right]\right| \leqq 2|\Delta'(\nu)||T_{nj}|$$

whenever

$$j \leqq \frac{n}{2}, \quad \frac{1}{n-j}|T_{nj}| \leqq \gamma, \quad n \geqq N.$$

Thus, for $j \leqq n/2$ and $n \geqq N$, $W_{nj} < r$ implies that either $|T_{nj}| > \gamma n/2$ or $|T_{nj}| > (j\Delta(y) - r)/2|\Delta'(\nu)|$ and, if $\delta > 0$ is sufficiently small, then the former event is a subset of the latter for all $j \leqq n\delta$. Next, since $y = y_a \to \nu$, there are a_0, k_0 and $\eta_0 > 0$ for which

$$\frac{j\Delta(y) - r}{2|\Delta(\nu)|j} - |y - \nu| > \eta_0$$

for all $j > k_0$ and $a > a_0$, for $j > k_0$ and $a > a_0$, $|T_{nj}| > (j\Delta(y) - r)/2|\Delta'(\nu)|$ implies that $|\bar{Y}_j - \nu| > \eta_0$. Thus, for sufficiently large a, k and sufficiently small $\delta > 0$, one has

$$\varepsilon_{a,1}(k, \delta) \leqq P\left\{|T_{nj}| > \frac{j\Delta(y) - r}{2|\Delta'(\nu)|}, \exists j \in (k, n\delta] \middle| \bar{Y}_n = y\right\}$$
$$\leqq P\{|\bar{Y}_j - \nu| > \eta_0, \exists j \in (k, n\delta] | \bar{Y}_n = y\}$$
$$\leqq CP\{|\bar{Y}_j - \nu| > \eta_0, \exists j \in (k, n\delta]\} \leqq CP\{|\bar{Y}_j - \nu| > \eta_0, \exists j > k\}$$

by Lemma 5.1 (iii). The latter term is independent of a and tends to zero as $k \to \infty$ by the S.L.L.N.

Remarks and references. Theorem 5.1 is adapted from Woodroofe (1976a) and Hagwood (1980). The possibility of computing $P\{t \geqq n | \bar{Y}_n = y\}$ by "looking backward" was suggested by Anscombe (1953).

Takahashi and Woodroofe (1981) have obtained asymptotic expansions in the special case that Y is normal and $\Delta(y) = y^2$.

5.2. Densities. In this section the limit of the joint mass function/density of t_a and R_a is obtained. As a corollary, t_a and R_a are shown to be asymptotically independent in a very strong sense. Recall that (under mild conditions)

$$t_a^* = \frac{t_a - a/\mu}{\sqrt{a/\mu}}$$

is asymptotically normal with mean 0 and variance $\mu^{-2}\sigma^2$, where $\mu=\Delta(\nu)$ and $\sigma^2=\Delta'(\nu)^2\tau^2$, and that R_a has an asymptotic distribution with density

$$h(r)=\frac{1}{\mu}P\{S_k \geqq r, k\geqq 1\},\qquad r>0.$$

See Theorem 2.7. Let k_a denote the joint mass function/density of t_a and R_a,

$$k_a(n,r)=\frac{d}{dr}P\{t_a=n, R_a\leqq r\},\qquad n\geqq 1,\quad r>0,$$

THEOREM. 5.2. *Suppose that condition* S *is satisfied, that* Δ *is strictly convex, that* Δ *is continuously differentiable near* ν *and that* $\Delta'(\nu)>0$. *If*

$$n=n_a=\frac{a}{\mu}+z_a\sqrt{\frac{a}{\mu}},\quad \text{where } z_a\to z, \tag{5.10}$$

then

$$k_a(n,r)\sim\frac{\mu}{\sigma\sqrt{n}}\Phi'\left(\frac{\mu z}{\sigma}\right)h(r)$$

as $a\to\infty$ *for all* $r>0$, *where* Φ' *denotes the standard normal density.*

Proof. Let $C_n=C_n(a,r)$ be the set of y for which $a<n\Delta(y)\leqq a+r$. Then, for sufficiently large n,

$$P\{t_a=n, R_a\leqq r\}=P\{t_a\geqq n, \bar{Y}_n\in C_n\}=\int_{C_n}u_a(n,y)ng_n(ny)\,dy,$$

where $u_a(n,y)=P\{t_a\geqq n\,|\,\bar{Y}_n=y\}$ and g_n denotes the density of $n\bar{Y}_n$. It follows that

$$k_a(n,r)=\sum_{y:n\Delta(y)=a+r}u_a(n,y)g_n(ny)\left|\frac{1}{\Delta'(y)}\right| \tag{5.11}$$

for all sufficiently large a.

Since Δ is strictly convex, there are at most two solutions to the equation $n\Delta(y)=a+r$ for fixed a, n and r. In fact, there is one solution $y_0=y_0(a,n,r)$ for which

$$y_0=\nu-\frac{1}{\sqrt{n}}z_a\frac{\Delta(\nu)}{\Delta'(\nu)}+o\left(\frac{1}{\sqrt{n}}\right)\quad \text{as } a\to\infty, \tag{5.12}$$

and the other solution, if any, is bounded away from ν. For if $\varepsilon>0$ is so small that $\Delta'(y)>0$ for $\nu-\varepsilon\leqq y\leqq\nu+\varepsilon$, then there can be at most one solution in the interval $[\nu-\varepsilon,\nu+\varepsilon]$, and since $n\Delta(\nu-\varepsilon)<a+r<n\Delta(\nu+\varepsilon)$ for all sufficiently large a, under the limiting operation (5.10), there must be at least one solution y_0 for sufficiently large a. Finally, (5.12) follows by expanding $a+r=n\Delta(y_0)$ in a Taylor series about $y_0=\nu$ and solving for y_0.

Theorem 5.2 then follows by substituting (5.12) into (5.11) and using Theorem 5.1 and the local limit theorem.

Observe that global convexity of Δ was not crucial to the proof. If Δ were only assumed to be strictly convex near ν, then Theorem 5.2 would hold with k_a replaced by $k_a^*(n, r) = (d/dr)P\{t_a = n, R_a \leqq r, |\bar{Y}_n - \nu| \leqq \varepsilon\}$ for small $\varepsilon > 0$, and $P\{|\bar{Y}_n - \nu| > \varepsilon\} = O(1/n)$ for any $\varepsilon > 0$.

COROLLARY 5.2. *Let h_a denote the marginal density of R_a, $a > 0$. If Δ is twice continuously differentiable near Δ, then*

$$\int_0^\infty |h_a(r) - h(r)|\, dr \to 0 \quad \text{as } a \to \infty.$$

COROLLARY 5.3. *Suppose that $E[\Delta(Y_1)^+] < \infty$. If (5.10) holds, then*

$$P\{t_a = n\} \sim \frac{\mu}{\sigma\sqrt{n}} \Phi'\left(\frac{\mu z}{\sigma}\right) \quad \text{as } a \to \infty.$$

COROLLARY 5.4. *Suppose that $E[\Delta(Y_1)^+] < \infty$ and let $h_a(\cdot|n)$ denote the conditional density of R_a, given $t_a = n$. If (5.10) holds, then*

$$h_a(r|n) \to h(r), \qquad r > 0, \quad a \to \infty.$$

Proofs. For Corollary 5.2, let $N_a = [a/\mu]$ and write

$$h_a(r) = \left\{ \sum_{|n - N_a| \leqq c\sqrt{N_a}} + \sum_{|n - N_a| > c\sqrt{N_a}} \right\} k_a(n, r) = h_{a,1}(r) + h_{a,2}(r),$$

say, for $a, r > 0$. Now, for each fixed $c > 0$, the limit in Theorem 5.2 holds uniformly in n for which $|n - N_a| \leqq c\sqrt{N_a}$, so that $h_{a,1}(r) \to h(r)[\Phi(\mu c/\sigma) - \Phi(-\mu c/\sigma)]$ for $c > 0$. Thus, $h_{a,1}(r) \to h(r)$ for some function $c = c_a \to \infty$ as $a \to \infty$. Moreover, with $c = c_a \to \infty$,

$$\int_0^\infty h_{a,2}(r)\, dr = P\{|t_a^*| > c_a\} \to 0 \tag{5.13}$$

as $a \to \infty$ by Lemma 4.2. Since (5.13) implies that $\int_0^\infty h_{a,1}(r)\, dr \to 1 = \int_0^\infty h(r)\, dr$, Corollary 5.2 now follows from Scheffe's theorem.

The proof of Corollary 5.3 is similar. For $c > 0$,

$$\sqrt{n}P\{t_a = n\} = \sqrt{n}\left\{\int_0^c + \int_c^\infty\right\} k_a(n, r)\, dr. \tag{5.14}$$

For each fixed $c > 0$, the first integral on the right side of (5.14) converges to $H(c)(\mu/\sigma)\Phi'(\mu z/\sigma)$ by Theorem 5.2 and the bounded convergence theorem; and $H(c) \to 1$ as $c \to \infty$. So it suffices to show that the second integral in (5.14) converges to zero as first $a \to \infty$ and then $c \to \infty$. Now

$$\sqrt{n}\int_c^\infty k_a(n, r)\, dr = \sqrt{n}P\{t_a = n, R_a > c\} \leqq \sqrt{n}P\{Z_{n-1} \leqq a, \Delta(Y_n) > c + a - Z_{n-1}\},$$

since the convexity of Δ implies $Z_n - Z_{n-1} \leqq \Delta(Y_n)$. Let $\varepsilon > 0$ be so small that $\Delta'(y) > 0$ for $\nu - \varepsilon \leqq y \leqq \nu + \varepsilon$, and let L denote the distribution of $\Delta(Y_1)$. Then

$P\{|\bar{Y}_n - \nu| > \varepsilon\} = O(1/n)$ as $a \to \infty$, and there is a $B > 0$ for which

$$\sqrt{n}P\{a + c - w \leqq Z_{n-1} \leqq a, |\bar{Y}_{n-1} - \nu| \leqq \varepsilon\} \leqq B(w - c), \qquad w > c,$$

so

$$\begin{aligned} &\sqrt{n}P\{Z_{n-1} \leqq a, \Delta(Y_n) > a + c - Z_{n-1}\} \\ &\quad \leqq \int_c^\infty \sqrt{n}P\{a + c - w \leqq Z_{n-1} \leqq a, |\bar{Y}_{n-1} - \nu| \leqq \varepsilon\} L\{dw\} + o(1) \\ &\quad \leqq B \int_c^\infty [1 - L(w)]\, dw + o(1) \quad \text{as } a \to \infty. \end{aligned}$$

Corollary 5.3 follows, since the last integral tends to zero as $c \to \infty$.

Corollary 5.4 follows directly from Corollary 5.3 and Theorem 5.2. When combined with Scheffe's theorem. Corollary 5.4 asserts that the conditional distribution of R_a converges strongly to H. This is a much stronger assertion of asymptotic independence than that of Theorem 4.2.

Remark. I believe the formulation of Theorem 5.2 and Corollary 5.4 to be new, at least in this generality.

CHAPTER 6

Open-ended Tests

6.1. In exponential families. Let Ω be a nondegenerate interval and let G_ω, $\omega \in \Omega$, be a one-parameter, exponential family with natural parameter space Ω; that is,

$$G_\omega\{dy\} = \exp[\omega y - \psi(\omega)]\lambda\{dy\}, \qquad -\infty < y < \infty, \quad \omega \in \Omega, \tag{6.1}$$

for some nondegenerate, sigma-finite measure λ, and Ω consists of all ω for which $\exp(\omega y)$ is integrable with respect to λ. Next, let $Y_1, Y_2, \cdots$ be i.i.d. with common distribution G_ω for some unknown $\omega \in \Omega$, let $\omega_0 \in \Omega^0$, the interior of Ω, and consider the hypotheses $\omega = \omega_0$ versus $\omega \neq \omega_0$. This chapter develops a class of sequential tests which have power one against all alternatives $\omega \neq \omega_0$. The price paid for this remarkable property is that the sample size is infinite with positive probability, when $\omega = \omega_0$.

The tests depend on two design parameters, a mixing distribution π over Ω and a critical level $c > 1$. To describe them, let $\mathcal{A}_n = \sigma\{Y_1, \cdots, Y_n\}$, $n \geqq 1$, $\mathcal{A}_\infty = \sigma\{Y_1, Y_2, \cdots\}$, and let P_ω be the unique probability measure on $\mathcal{A}_\infty$ under which $Y_1, Y_2, \cdots$ are i.i.d. with common distribution G_ω, $\omega \in \Omega$. Next, given a mixing distribution π over Ω, define Q by

$$Q(A) = \int_\Omega P_\omega(A)\pi\{d\omega\}, \qquad A \in \mathcal{A}_\infty. \tag{6.2}$$

Then Q is easily seen to be a probability measure. Moreover, letting P_ω^n and Q^n denote the restrictions of P_ω and Q to $\mathcal{A}_n$, one finds

$$L_n = \frac{dQ^n}{dP_{\omega_0}^n} = \int_\Omega \exp\{(\omega - \omega_0)n\bar{Y}_n - n[\psi(\omega) - \psi(\omega_0)]\}\pi\{d\omega\}, \qquad n \geqq 1, \tag{6.3}$$

by reversing the order of integration in (6.2). If $c > 1$, then the open-ended test, with mixing distribution π and critical level c, continues to observe $Y_1, Y_2, \cdots$ as long as $L_n \leqq c$; if $L_n > c$ for some $n \geqq 1$, then the test stops with the first such n and rejects H_0; if $L_n \leqq c$ for all $n \geqq 1$, then sampling continues indefinitely and H_0 is never rejected. Alternatively, the test rejects H_0 if and only if the stopping time

$$t = \inf\{n \geqq 1 : L_n > c\}$$

is finite (where $\inf \emptyset = \infty$). Properties of the test are detailed in Theorems 6.1–6.3.

Recall that the support of a measure μ, defined on Borel sets of $(-\infty, \infty)$, consists of all x for which $\mu\{(x-\varepsilon, x+\varepsilon)\}>0$ for all $\varepsilon>0$. Alternatively, the support is the smallest closed set C for which $\mu\{C'\}=0$.

THEOREM 6.1. *For any mixing distribution π and any $c>1$,*

$$P_{\omega_0}\{t<\infty\}\leqq\frac{1}{c}.$$

Moreover,

$$P_\omega\{t<\infty\}=1 \quad \textit{for all } \omega\neq\omega_0,$$

if the support of π includes all of Ω.

Proof. Since t is a stopping time with respect to $\mathcal{A}_n$, $n\geqq 1$, it follows from Theorem 1.1 that

$$P_{\omega_0}\{t<\infty\}=\int_{t<\infty}\left(\frac{1}{L_t}\right)dQ\leqq\frac{1}{c}Q\{t<\infty\}\leqq\frac{1}{c}.$$

This establishes the first assertion. The second is established by showing that $L_n\to\infty$ w.p. 1 (P_ω) as $n\to\infty$ for all $\omega\neq\omega_0$, in Lemma 6.1 below.

Example 6.1. (i) If G_ω is the normal distribution with unknown mean ω, $-\infty<\omega<\infty$, and unit variance, then G_ω are of the form (6.1) with $\lambda=G_0$ and $\psi(\omega)=\omega^2/2$, $-\infty<\omega<\infty$. If $\omega_0=0$ and if π is the normal distribution with mean 0 and variance $1/r$, then

$$L_n=\sqrt{\frac{r}{n+r}}\exp\left[\frac{(n\bar{Y}_n)^2}{2(n+r)}\right], \qquad n\geqq 1,$$

by a direct computation. For example, when $r=1$, t is the first n, if any, for which $|n\bar{Y}_n|>\{2(n+1)[\log c+\frac{1}{2}\log(n+1)]\}^{1/2}$.

(ii) If G_ω is the exponential distribution with failure rate $|\omega|$, $-\infty<\omega<0$, then G_ω are of the form (6.1) with $\lambda=$ Lebesgue measure over $(0,\infty)$ and $\psi(\omega)=-\log|\omega|$, $-\infty<\omega<0$. If $\omega_0=-1$ and if π is the exponential distribution with failure rate $\beta>0$, then

$$L_n=\frac{\beta n!}{(\beta+n\bar{Y}_n)^{n+1}}\exp(n\bar{Y}_n), \qquad n\geqq 1.$$

The following lemmas develop some asymptotic properties of the likelihood ratios L_n, $n\geqq 1$. In them, it is assumed that $\omega_0=0$ and $\psi(0)=0$. This assumption does not limit the generality, since it may be achieved by a reparameterization.

Observe first that $\psi''(\omega)=D_\omega(Y_1)>0$, $\omega\in\Omega^0$, since G_ω is nondegenerate, $\omega\in\Omega$. Thus ψ is a strictly convex function on Ω, and ψ' is a strictly increasing function from Ω^0 onto its range $\mathcal{Y}_0=\psi'(\Omega^0)$, which is called the expectation space. Next, let

$$\phi(y)=\sup_{\omega\in\Omega}[\omega y-\psi(\omega)], \qquad -\infty<y<\infty. \tag{6.4}$$

Then ϕ is an extended-valued, convex function on $(-\infty,\infty)$. The function ϕ is finite and twice continuously differentiable on $\mathcal{Y}_0=\psi'(\Omega^0)$, for if $y\in\mathcal{Y}_0$, then the supremum in (6.4) is attained at $\hat{\omega}=\hat{\omega}(y)$, where

$$\psi'(\hat{\omega})=y, \qquad \phi(y)=\hat{\omega}y-\psi(\hat{\omega}).$$

The function $\hat{\omega}$ is continuously differentiable on $\mathcal{Y}_0$ by the implicit function theorem, so $\phi'=\hat{\omega}$ is continuously differentiable on $\mathcal{Y}_0$, too. The functions $\hat{\omega}$ and ϕ have direct statistical interpretations. Given $Y_1,\cdots,Y_n$, the log-likelihood function is $l_n(\omega;Y_1,\cdots,Y_n)=n(\omega\bar{Y}_n-\psi(\omega))$, $\omega\in\Omega$; if $\bar{Y}_n\in\mathcal{Y}_0$, then $\hat{\omega}(\bar{Y}_n)$ is the maximum likelihood estimate of ω, and $\Lambda_n=n\phi(\bar{Y}_n)$ is the log-likelihood ratio statistic for testing H_0: $\omega=0$.

LEMMA 6.1. *If support* $(\pi)=\Omega$, *then*

$$\frac{1}{n}\log\int_\Omega \exp\{n[sy-\psi(s)]\}\pi\{ds\}\uparrow\phi(y) \tag{6.5}$$

uniformly in y on compact subintervals of $\mathcal{Y}_0$ *as* $n\to\infty$, *and* $(1/n)\log L_n\to\phi[\psi'(\omega)]$ w.p. 1 (P_ω) *as* $n\to\infty$ *for all* $\omega\in\Omega^0$.

Proof. It is well known that (6.5) holds for each fixed $y\in\mathcal{Y}_0$ (see, for example, Royden (1968, p. 112)), and since both sides of (6.5) are continuous in y, uniformity on compact intervals follows from Dini's theorem. Since $\bar{Y}_n\to\psi'(\omega)$ w.p. 1 (P_ω) as $n\to\infty$ for all $\omega\in\Omega^0$, the second assertion follows directly from the first.

Since $\phi(y)>0$ for $0\neq y\in\mathcal{Y}_0$, the lemma implies that $L_n\to\infty$ with probability 1 (P_ω) as $n\to\infty$ for all $\omega\in\Omega^0$ with $\omega\neq 0=\omega_0$, as required in Theorem 1. The endpoints, if any, may be handled similarly.

In the second lemma, relation (6.5) is refined. Let

$$v_n(y)=\exp[-n\phi(y)]\int_\Omega \exp\{n[sy-\psi(s)]\}\pi\{ds\}, \qquad y\in\mathcal{Y}_0. \tag{6.6}$$

LEMMA 6.2. *Suppose that* π *has a positive continuous density* ρ *with respect to Lebesgue measure on* Ω. *Then*

$$v_n(y)\sim\sqrt{\frac{2\pi}{n\psi''(\hat{\omega})}}\,\rho(\hat{\omega})$$

uniformly in y on compact subintervals of $\mathcal{Y}_0$ *as* $n\to\infty$.

Proof. Let $H(\omega,y)=[\psi(\omega)-\psi(\hat{\omega})]-\psi'(\hat{\omega})(\omega-\hat{\omega})$ for $\omega\in\Omega^0$ and $y\in\mathcal{Y}_0$. Then

$$v_n(y)=\int_\Omega \exp[-nH(s,y)]\rho(s)\,ds, \qquad y\in\mathcal{Y}_0.$$

Observe that $H(\omega,y)$ is convex in ω for fixed $y\in\mathcal{Y}_0$, since ψ is convex. Moreover, for fixed y, $H(\omega,y)=\frac{1}{2}\psi''(\omega^*)(\omega-\hat{\omega})^2$, where $\omega^*=\omega^*(\omega,y)$ is an intermediate point between ω and $\hat{\omega}$. Let K be any compact subinterval of $\mathcal{Y}_0$. Then there are a $\sigma>0$ and a compact $J\subset\Omega^0$ for which $[\hat{\omega}-\delta,\hat{\omega}+\delta]\subset J$ for all $y\in K$ and, since ψ'' is positive and continuous, there is an $\varepsilon>0$ for which $\psi''(\omega^*)\geqq\varepsilon$ for

$|\omega-\hat{\omega}|\leqq\delta$ and $y\in K$. In particular, $H(\omega, y)\geqq\frac{1}{2}\varepsilon(\omega-\hat{\omega})^2$ for $|\omega-\hat{\omega}|\leqq\delta$ and $y\in K$. Since H is convex in ω for fixed y, it follows that $H(\omega, y)\geqq\frac{1}{2}\varepsilon\delta^2$ for $|\omega-\hat{\omega}|\geqq\delta$ and $y\in K$ and, consequently, that

$$\int_{|\omega-\hat{\omega}|\geqq\delta} e^{-nH}\,d\pi\leqq e^{-\varepsilon\delta^2 n/2}, \qquad y\in K,\quad n\geqq 1.$$

Next, let $y_n\in K$, $n\geqq 1$, be a convergent sequence, say $y_n\to y\in K$ as $n\to\infty$, let $\hat{\omega}_n=\hat{\omega}(y_n)$, $n\geqq 1$ and $\hat{\omega}=\hat{\omega}(y)$, and consider the integral over $|\omega-\hat{\omega}_n|<\delta$. The change of variables $\hat{\omega}=\omega_n+n^{-1/2}s$ shows that

$$\text{(6.7)}\qquad \sqrt{n}\int_{|\omega-\hat{\omega}_n|<\delta} e^{-nH}\,d\pi=\int_{-\delta\sqrt{n}}^{\delta\sqrt{n}}\exp\left[-\frac{1}{2}\psi''(\omega_n^*)s^2\right]\rho\left(\hat{\omega}_n+\frac{s}{\sqrt{n}}\right)ds$$

where $\omega_n^*=\omega^*(\hat{\omega}_n+n^{-1/2}s, y_n)$, $n\geqq 1$. As $n\to\infty$, the integrand on the right side of (6.7) converges to $\exp[-\frac{1}{2}\psi''(\hat{\omega})s^2]\rho(\hat{\omega})$; and the integrand is dominated by $C\exp(-\frac{1}{2}\varepsilon s^2)$ for some C. So, the right side of (6.7) converges to

$$\int_{-\infty}^{\infty}\exp\left[-\frac{1}{2}\psi''(\hat{\omega})s^2\right]\rho(\hat{\omega})\,ds=\sqrt{\frac{2\pi}{\psi''(\hat{\omega})}}\rho(\hat{\omega})$$

by the dominated convergence theorem. The lemma follows.

Remarks and references. Power one tests and the related theory of confidence sequences were developed by Darling and Robbins (1967a, b, c), (1968) and Robbins and Siegmund (1970). Much of this research is described in Robbins' (1970) Wald Lectures.

6.2. Error probabilities. Nonlinear renewal theory may be used to refine the inequality $P_{\omega_0}\{t<\infty\}\leqq 1/c$ of Theorem 6.1. To see how, let π be a prior distribution; let $c>1$ and let $L_n=dQ^n/dP^n_{\omega_0}$, $n\geqq 1$, denote the likelihood ratios. Also, let

$$Z_n=\log L_n,\qquad n\geqq 1,$$

$$a=\log c,$$

$$t=t_a=\inf\{n\geqq 1\colon Z_n>a\},$$

$$R_a=Z_{t_a}-a\quad\text{ on }\{t_a<\infty\},\quad a>0.$$

Then the open-ended test rejects H_0: $\omega=\omega_0$ if and only if $t_a<\infty$. If $1<N\leqq\infty$, then

$$\text{(6.8)}\qquad P_{\omega_0}\{t_a<N\}=\int_{t_a<N}\exp(-Z_{t_a})\,dQ=\mathcal{I}(a, N)\,e^{-a}$$

where

$$\mathcal{I}(a, N)=\int_{t_a<N}\exp(-R_a)\,dQ=\int_{\Omega}\left[\int_{t_a<N}\exp(-R_a)\,dP_\omega\right]\pi\{d\omega\}.$$

Nonlinear renewal theory is used to approximate the inner integral for fixed ω and the approximations are then integrated.

To so apply the nonlinear renewal theory, one needs to write $Z_n = S_n + \xi_n$, $n \geqq 1$, where $S_n = S(\omega)$, $n \geqq 1$, is a random walk and $\xi_n = \xi_n(\omega)$, $n \geqq 1$ are slowly changing, both with respect to P_ω. For fixed $\omega \neq \omega_0$,

$$Z_n = S_n + \xi_n, \qquad n \geqq 1,$$

where

$$S_n = S_n(\omega) = (\omega - \omega_0) n \bar{Y}_n - n[\psi(\omega) - \psi(\omega_0)], \qquad n \geqq 1,$$

and

$$\xi_n = \xi_n(\omega) = \log \int_\Omega \exp\{(s-\omega) n \bar{Y}_n - n[\psi(s) - \psi(\omega)]\} \pi\{ds\}, \qquad n \geqq 1.$$

Clearly, S_n, $n \geqq 1$, is a random walk (with respect to P_ω) with summands $X_i = X_i(\omega) = (\omega - \omega_0) Y_i - [\psi(\omega) - \psi(\omega_0)]$, $i \geqq 1$. Observe that $X_1 = X_1(\omega)$ is the log-likelihood ratio $X_1 = \log dP^1_\omega / dP^1_{\omega_0}$. Thus, the mean of X_1 is the Kullback–Leibler information measure

$$J(\omega_0, \omega) = E_\omega(X_1) = (\omega - \omega_0)\psi'(\omega) - [\psi(\omega) - \psi(\omega_0)],$$

which is positive if $\omega \neq \omega_0$.

LEMMA 6.3. *Suppose that π has a positive continuous density ρ with respect to Lebesgue measure on Ω. Then ξ_n, $n \geqq 1$, are slowly changing with respect to P_ω for any $\omega \in \Omega^0$.*

Proof. Fix $\omega \in \Omega^0$ and recall that $\mathcal{Y}_0 = \psi'(\Omega^0)$. For $n \geqq 1$, let B_n be the event that $\bar{Y}_k \in \mathcal{Y}_0$ for all $k \geqq n$. Then $I_{B_n} \to 1$ w.p. 1 (P_ω) as $n \to \infty$, since $\bar{Y}_n \to \psi'(\omega) \in \Omega^0$. So, it suffices to show that $\xi_n I_{B_n}$, $n \geqq 1$, are slowly changing. If B_n occurs, then $\hat{\omega}_n = \hat{\omega}(\bar{Y}_n)$ is well defined and $\xi_n = \xi'_n + \xi''_n$, where

$$\xi'_n = (\hat{\omega}_n - \omega) n \bar{Y}_n - n[\psi(\hat{\omega}_n) - \psi(\omega)] = nJ(\omega, \hat{\omega}_n),$$

$$\xi''_n = \log \int_\Omega \exp\{(s - \hat{\omega}_n) n \bar{Y}_n - n[\psi(s) - \psi(\hat{\omega}_n)]\} \pi\{ds\}, \qquad n \geqq 1.$$

Now

$$\xi''_n = -\tfrac{1}{2} \log n + \log u_n(\bar{Y}_n),$$

where $u_n(y)$ converges to a finite positive limit, uniformly in y on compact subintervals of $\mathcal{Y}_0$, by Lemma 6.2. It follows that $\xi''_n I_{B_n}$, $n \geqq 1$, are slowly changing. Next, a Taylor series expansion shows that

$$\xi'_n = \frac{1}{2\psi''(\omega)} n[\bar{Y}_n - \psi'(\omega)]^2 + C_n \cdot n[\bar{Y}_n - \psi'(\omega)]^3,$$

where C_n is convergent w.p. 1 (P_ω). The second term on the right converges to zero w.p. 1 (P_ω) by the law of the iterated logarithm and the first is slowly changing by Example 4.1.

LEMMA 6.4. *For a.e. ω (Lebesgue), $X_1(\omega)=(\omega-\omega_0)Y_1-[\psi(\omega)-\psi(\omega_0)]$ has a nonarithmetic distribution, when $Y_1 \sim G_\omega$.*

Proof. Suppose that $X_1(\omega)$ has an arithmetic distribution for some $\omega \neq \omega_0$, say $\omega=\omega_1$, and let Δ_1 be the span of $X_1(\omega_1)$. Then Y_1 must take values of the form

$$\left(\frac{1}{\omega_1-\omega_0}\right)\{k\Delta_1+[\psi(\omega_1)-\psi(\omega_0)]\}=\Delta k+\gamma, \quad \text{say},$$

where $k=0, \pm 1, \pm 2, \cdots$. Moreover, since Y_1 is assumed to have a nondegenerate distribution, there are $k_1 \neq k_2$ for which $\lambda\{\Delta k_1+\gamma\}>0<\lambda\{\Delta k_2+\gamma\}$. Now suppose that $X_1(\omega)$ has an arithmetic distribution for some ω with $\omega_0 \neq \omega \neq \omega_1$ and let $d(\omega)>0$ denote the span of $X_1(\omega)$. Then there are j_1 and j_2 for which $j_1 \neq j_2$ and

$$j_i d(\omega)=(\omega-\omega_0)(\Delta k_i+\gamma)-[\psi(\omega)-\psi(\omega_0)], \qquad i=1, 2.$$

So

$$\frac{\psi(\omega)-\psi(\omega_0)}{\omega-\omega_0}=\gamma+\Delta\left(\frac{k_1 j_2-k_2 j_1}{j_2-j_1}\right). \tag{6.9}$$

Thus, the set of ω for which $\omega_0 \neq \omega \neq \omega_1$ and $X_1(\omega)$ has an arithmetic distribution is contained in the set of ω for which (6.9) holds for some $j_1 \neq j_2$; the latter set is countable, since ψ is convex.

It follows from the lemma that for a.e. ω, R_a has limiting distribution

$$H_\omega\{dr\}=\frac{1}{E(S_\tau)}P\{S_\tau>r\}\,dr, \qquad r>0, \tag{6.10}$$

where $\tau=\tau(\omega)$ denotes the first strict ascending ladder epoch for the random walk $S_k=S_k(\omega)$, $k \geqq 0$. The Laplace transform of this distribution plays an important role below and is denoted by $\mathcal{H}(\omega, \cdot)$.

THEOREM 6.2. *Suppose that π has a positive continuous density with respect to Lebesgue measure on Ω. Suppose also that $N=N(a)\to\infty$ as $a\to\infty$ with $N/a\to 1/\delta$, where $0\leqq\delta<\infty$. Then*

$$P_{\omega_0}\{t_a<N\}\sim K e^{-a} \quad \textit{as } a\to\infty,$$

where

$$K=K(\omega_0, \pi, \delta)=\int_{J(\omega_0,\omega)>\delta}\mathcal{H}(\omega, 1)\pi\{d\omega\}$$

and $\mathcal{H}(\omega, \cdot)$ denotes the Laplace transform of the distribution (6.10).

Proof. From (6.8), it suffices to show that $\mathcal{I}(a, N)\to K$ as $a\to\infty$. Fix $\omega\neq\omega_0$. Then, since the mean of $X_1(\omega)$ is $J(\omega_0, \omega)>0$, $t_a/a\to 1/J(\omega_0, \omega)$ w.p. 1 as $a\to\infty$. Thus,

$$P_\omega\{t_a<N\}\to\begin{cases}0, & J(\omega_0, \omega)<\delta\\ 1, & J(\omega_0, \omega)>\delta\end{cases} \qquad \text{as } a\to\infty.$$

Observe also that $J(\omega_0, \omega) = \delta$ for at most two values of ω. Since R_a has the limiting distribution (6.10) for a.e. ω, it now follows that

$$(6.11) \qquad \int_{t_a<N} \exp(-R_a)\, dP_\omega \to \mathcal{H}(\omega, 1) I_{\{J(\omega_0,\omega)>\delta\}}$$

as $a \to \infty$ for a.e. ω. Finally, since the left side of (6.11) is bounded, the limit in (6.11) may be integrated with respect to π to complete the proof. See (6.8).

COROLLARY 6.1. *$P_{\omega_0}\{t_a < \infty\} \sim K e^{-a}$ as $a \to \infty$, where*

$$K = K(\omega_0, \pi) = \int \mathcal{H}(\omega, 1)\pi\{d\omega\}.$$

The constants K of Theorem 6.2 and its corollary are typically complicated. Their computation is illustrated in the examples below.

Example 6.2. *The normal case.* If G_ω is the normal distribution with mean ω, $-\infty < \omega < \infty$, and unit variance, and if $\omega_0 = 0$, then $X_1(\omega) = \omega Y_1 - \omega^2/2$ has the normal distribution with mean $\mu = \omega^2/2$ and variance ω^2, under P_ω for $\omega \neq 0$. By Example 3.1, the Laplace transform is

$$\mathcal{H}(\omega, 1) = 2\omega^{-2} \exp\left\{-2 \sum_{k=1}^{\infty} \frac{1}{k} \Phi\left(-\frac{1}{2}|\omega|\sqrt{k}\right)\right\}, \qquad \omega \neq 0.$$

The expression for $\mathcal{H}$ does not simplify, but it is amenable to numerical calculation. A brief table was included in Example 3.1. For example, when π is the standard normal distribution and $\delta = 0$, one finds $K \approx \frac{2}{3}$, so, the inequality of Theorem 6.1 may substantially overestimate $P_a\{t_a < \infty\}$. Table 6.1 below presents the results of a small Monte Carlo experiment to test the accuracy of the approximations. The asymptotic theory yields approximations which are consistently larger than the Monte Carlo values.

TABLE 6.1
The error probability $\alpha = P_0\{t \leqq a\delta^{-1}\}$

		Asymptotic theory	Monte Carlo ± S.D.
$c = 10$	$\delta = .04$	.0461	.0454 ± .0017
	$\delta = .02$	.0517	.0500 ± .0017
$c = 20$	$\delta = .04$	.0229	.0217 ± .0009
	$\delta = .02$	.0257	.0246 ± .0008
$c = 100$	$\delta = .04$	.00462	.00450 ± .00018
	$\delta = .02$	.00518	.05512 ± .00018

Here Y is normally distributed with unknown mean ω and unit variance $\omega_0 = 0$ and π is the standard normal distribution. The asymptotic values are computed from Theorem 6.2 with $a = \log c$. Source: Lai and Siegmund (1977).

Example 6.3. *The exponential case.* Suppose that G_ω is the exponential distribution with failure rate $|\omega|$, $-\infty<\omega<0$ and that $\omega_0=-1$. Then

$$X_1(\omega)=(1-|\omega|)Y_1+\log|\omega|,$$

$$J(-1,\omega)=\frac{1}{|\omega|}[|\omega|\log|\omega|-(|\omega|-1)],\qquad \omega\neq -1.$$

The Laplace transform $\mathcal{H}(\omega, 1)$ was computed in Example 3.2 as

$$\mathcal{H}(\omega,1)=\begin{cases}[(|\omega|-1)-\log|\omega|]/[|\omega|\log|\omega|-(|\omega|-1)], & \omega<-1,\\ |\omega|, & -1<\omega<0.\end{cases}$$

For example, when π is the standard exponential distribution and $\delta=0$, one finds that $K\approx .567$.

Remarks and references. Theorem 6.2 and Table 6.1 were taken from Lai and Siegmund (1977). The Monte Carlo experiments used importance sampling which is discussed by Siegmund (1976) in the context of the S.P.R.T. Takahashi (1978) has obtained bounds on the rate of convergence in the normal case.

6.3. The expected sample size. In this section an approximation to $E_\omega(t_a)$ is developed. The notation of § 6.2 is used throughout.

THEOREM 6.3. *Suppose that π has a positive continuous density q with respect to Lebesgue measure on Ω. If $\omega\in\Omega^0$, $\omega\neq\omega_0$, and $X_1(\omega)=(\omega-\omega_0)Y_1-[\psi(\omega)-\psi(\omega_0)]$ does not have an arithmetic distribution, then*

$$E_\omega(t_a)=\frac{1}{J}\left[a+\frac{1}{2}\log\frac{a}{J}+C(\omega)\right]+o(1)\quad \textit{as } a\to\infty, \tag{6.12}$$

where

$$C(\omega)=\rho_\omega-\frac{1}{2}\log\frac{2\pi}{\psi''(\omega)}-\log q(\omega)-\frac{1}{2},$$

$$J=J(\omega_0,\omega)=(\omega-\omega_0)\psi'(\omega)-[\psi(\omega)-\psi(\omega_0)]$$

denotes the Kullback–Leibler information number, and ρ_ω denotes the mean of the asymptotic distribution H_ω of (6.10).

Proof. Given $\omega\in\Omega^0$ with $\omega\neq\omega_0$, $Z_n=S_n+\xi_n$, $n\geqq 1$, where $\xi_n=\xi_n(\omega)$, $n\geqq 1$ are slowly changing and $S_n=S_n(\omega)$, $n\geqq 1$, is a random walk with drift $E_\omega[S_1(\omega)]=J$. It will be shown that ξ_n, $n\geqq 1$, satisfy the conditions (4.10)–(4.16) of Theorem 4.5. Relation (6.12) will then follow by specialization.

First, let $\varepsilon>0$ be so small that $|\omega-\omega_0|\geqq 2\varepsilon$ and $[\psi'(\omega)-\varepsilon,\psi'(\omega)+\varepsilon]\subset\mathcal{Y}_0=\psi'(\Omega^0)$, and let A_n be the event that $|\bar{Y}_n-\psi'(\omega)|\leqq\varepsilon$. Then there is a $\delta>0$ for which $P(A_n')\leqq\exp(-\delta n)$, $n\geqq 1$, since Y_1 has a finite moment generating function. So condition (4.10) is satisfied. If A_n occurs, then the maximum likelihood estimate $\hat{\omega}_n=\hat{\omega}(\bar{Y}_n)$ is well defined and

$$\xi_n=l_n+V_n,$$

where

$$l_n = -\tfrac{1}{2}\log n$$

and

$$V_n = nJ(\omega, \hat{\omega}_n) + \log u_n(\bar{Y}_n)$$

and $u_n(y) \to \sqrt{[2\pi/\psi''(\hat{\omega})]}q(\hat{\omega})$ uniformly in y on compact subsets of $\mathcal{Y}_0$. See the proof of Theorem 6.2. Clearly, $l_n = -\frac{1}{2}\log n$ satisfy condition (4.12). And

$$V_n \Rightarrow \frac{1}{2}\chi_1^2 + \frac{1}{2}\log\frac{2\pi}{\psi''(\omega)} + \log q(\omega),$$

where χ_1^2 denotes a random variable having the chi-squared distribution on one degree of freedom, so that (4.15) is satisfied. Next, since $u_n(\bar{Y}_n)$ is bounded on A_n, it suffices to show that $nJ(\omega, \hat{\omega}_n)$ satisfies (4.13) and (4.14). The latter is clearly satisfied, since $nJ(\omega, \hat{\omega}_n) \geqq 0$. To check the former, one may use Taylor's theorem and the fact that ψ'' is bounded away from zero on compact subsets of Ω to conclude that $nJ(\omega, \hat{\omega}_n) \leqq Bn[\bar{Y}_n - \psi'(\omega)]^2$ on A_n for all $n \geqq 1$ for some constant B. The verification then proceeds as in Example 4.4.

Finally, it is shown that $P_\omega\{t_a \leqq a/2J\} = o(1/a)$ as $a \to \infty$, thus verifying (4.16). Let $m = m(a) = [a/2J]$. Let $\varepsilon > 0$ be so small that $|\omega - \omega_0| \geqq 2\varepsilon$, $[\psi'(\omega) - \varepsilon, \psi'(\omega) + \varepsilon] \subset \mathcal{Y}_0$, as above, and $\varepsilon(\omega - \omega_0) < J/2$; and let $B = B_a$ be the event that $|\bar{Y}_m - \psi'(\omega)| \leqq \varepsilon$. Then $P(B_a') = o(1/a)$ as $a \to \infty$. If B occurs,

$$\begin{aligned}\frac{dP_\omega^m}{dP_{\omega_0}^m} &= \exp\{(\omega - \omega_0)m\bar{Y}_m - m[\psi(\omega) - \psi(\omega_0)]\}\\
&\leqq \exp\{mJ(\omega_0, \omega) + m\varepsilon(\omega - \omega_0)\}\\
&\leqq \exp\left(\frac{a}{2} + \frac{a}{4}\right) \leqq \exp\left(\frac{3a}{4}\right).\end{aligned}$$

So

$$P_\omega\{t_a \leqq m, B\} \leqq \exp\left(\frac{3a}{4}\right)P_{\omega_0}\{t_a \leqq m\} \leqq \exp\left(-\frac{a}{4}\right)$$

and

$$P_\omega\{t_a \leqq m\} \leqq P_\omega\{t_a \leqq m, B\} + P_\omega(B') = o\left(\frac{1}{a}\right) \quad \text{as } a \to \infty.$$

Example 6.4. *The normal case.* If G_ω is the normal distribution with unknown mean ω, $-\infty < \omega < \infty$, and if $\omega_0 = 0$, then $X_1(\omega) = \omega Y_1 - \omega^2/2$. The mean of the asymptotic distribution of residual waiting time for the random walk $S_k(\omega)$, $k \geqq 1$, was computed in Example 3.1 as

$$\rho_\omega = 1 + \frac{\omega^2}{4} - \sum_{k=1}^{\infty}\frac{1}{\sqrt{k}}\left[\Phi'\left(\frac{\omega}{2}\sqrt{k}\right) - \frac{\omega}{2}\sqrt{k}\Phi\left(-\frac{\omega}{2}\sqrt{k}\right)\right], \qquad \omega > 0.$$

and a brief table of ρ_ω was also included in Example 3.1. Thus the approximation (6.12) may be simply computed. Table 6.2 below compares the approximation

TABLE 6.2

The expected sample size $E_\omega(t)$

		Asymptotic theory	Monte Carlo ± S.D.
$\omega = .3$	$c = 20$	106.0	109.3 ± 2.4
	$c = 100$	146.6	153.3 ± 5.4
$\omega = .5$	$c = 20$	36.3	37.3 ± 1.1
	$c = 100$	50.9	51.6 ± 2.5
$\omega = .75$	$c = 20$	15.8	16.5 ± 0.6
	$c = 100$	22.3	22.7 ± 1.4

Here Y is normally distributed with unknown mean ω and unit variance, $\omega_0 = 0$ and π is the standard normal distribution. The asymptotic value is computed from Theorem 6.3 with $a = \log c$. The Monte Carlo values are taken from Pollak and Siegmund (1975).

(6.12) with the results of a Monte Carlo experiment. The approximation is consistently smaller than the Monte Carlo values by between one and two standard deviations.

Remarks and references. Theorem 6.3 is adapted from Pollak and Siegmund (1975), who give the expansion (6.12) neglecting the excess over the boundary. The simulations of Table 6.2 are also taken from Pollak and Siegmund (1975).

CHAPTER 7

Repeated Significance Tests

7.1. Likelihood ratio tests. Let G_ω, $\omega \in \Omega$, denote a one-parameter exponential family with natural parameter space Ω, as in Chapter 6. Thus

$$G_\omega\{dy\} = \exp[\omega y - \psi(\omega)]\lambda\{dy\}, \qquad -\infty < y < \infty, \quad \omega \in \Omega,$$

where λ is a nondegenerate, sigma-finite measure over $(-\infty, \infty)$ and Ω consists of all ω for which $\exp(\omega y)$ is integrable with respect to λ. In addition, it is now assumed that Ω is open, say $\Omega = (\underline{\omega}, \bar{\omega})$, where $-\infty \leqq \underline{\omega} < \bar{\omega} \leqq \infty$. Next, let $Y_1, Y_2, \cdots$ be i.i.d. with common distribution G_ω for some unknown $\omega \in \Omega$, and, given $\omega_0 \in \Omega$, consider testing the sharp hypothesis $\omega = \omega_0$ versus $\omega \neq \omega_0$, and the one-sided analogue $\omega \leqq \omega_0$ versus $\omega > \omega_0$.

Recall that the log-likelihood function, given $Y_1, \cdots, Y_n$, is $l_n(\omega) = n[\omega \bar{Y}_n - \psi(\omega)]$, $\omega \in \Omega$. Let

$$\phi(y) = \sup_{\Omega}\{(\omega - \omega_0)y - [\psi(\omega) - \psi(\omega_0)]\}, \qquad -\infty < y < \infty.$$

Then

$$\Lambda_n = n\phi(\bar{Y}_n)$$

is the log-likelihood ratio statistic for testing $\omega = \omega_0$ versus $\omega \neq \omega_0$ on the basis of $Y_1, \cdots, Y_n$, $n \geqq 1$. The function ϕ may be infinite for some values of y, but Λ_n is finite w.p. 1, as shown below. In particular, if $y \in \mathcal{Y}_0 = \psi'(\Omega)$, then $\phi(y) = J[\omega_0, \hat{\omega}]$, where $\psi'(\hat{\omega}) = y$ and J denotes the Kullback–Leibler information, $J(\omega_1, \omega_2) = (\omega_2 - \omega_1)\psi'(\omega_2) - [\psi(\omega_2) - \psi(\omega_1)]$ for $\omega_1, \omega_2 \in \Omega$; see § 6.1.

It is straightforward to compute the function ϕ in special cases.

Example 7.1. (i) If G_ω is the normal distribution with mean ω, $-\infty < \omega < \infty$, then the mean is $\theta = \psi'(\omega) = \omega$, $-\infty < \omega < \infty$. It follows easily that $\mathcal{Y}_0 = (-\infty, \infty)$ and that $\hat{\omega}(y) = y$, $-\infty < y < \infty$. If the null hypothesis is that $\omega = 0$, then

$$\phi(y) = \frac{y^2}{2}, \qquad -\infty < y < \infty.$$

(ii) If G_ω is the exponential distribution with failure rate $|\omega|$, $-\infty < \omega < 0$, then the mean is $\theta = \psi'(\omega) = 1/|\omega|$, $-\infty < \omega < 0$. It follows easily that $\mathcal{Y}_0 = (0, \infty)$ and that $\hat{\omega}(y) = -1/y$, $0 < y < \infty$. If the null hypothesis is $\omega = -1$, then

$$\phi(y) = (y - 1) - \log y, \qquad y > 0.$$

(iii) If G_ω is the Bernoulli distribution with mean $\theta = e^\omega/(e^\omega + 1)$, $-\infty < \omega < \infty$, then $\mathcal{Y}_0 = (0, 1)$ is the unit interval and $\hat{\omega}(y) = \log(y/(1-y))$ for $0 < y < 1$. If the null hypothesis is $\theta=\frac{1}{2}$, then

$$\phi(y)=y\log y+(1-y)\log(1-y)+\log 2, \qquad 0<y<1.$$

(iv) If G_ω is the Poisson distribution with mean $\theta=e^\omega$, $-\infty<\omega<\infty$, then $\mathcal{Y}_0=(0,\infty)$ and $\hat{\omega}(y)=\log y$ for $y>0$. If the null hypothesis is $\theta=1$, then

$$\phi(y)=y\log y-y+1, \qquad y>0.$$

The reader may recall that the null distribution of the likelihood ratio statistic Λ_n converges to a chi-squared distribution on one degree of freedom.

Given a sample of fixed size, say $Y_1,\cdots,Y_n$, the likelihood ratio test with critical level $c=e^a$ rejects the null hypothesis $\omega=\omega_0$ if and only if $\Lambda_n>a$. The probability of a type I error is then

$$\alpha=P_{\omega_0}\{\Lambda_n>a\},$$

which may often be estimated by using the chi-squared approximation to the null distribution of Λ_n.

By contrast, one may test $\omega=\omega_0$ repeatedly over time. If $1\leqq m<N$ are integers and $0<b\leqq a$, then *the repeated likelihood ratio test with initial sample size m, maximum sample size N, and (logarithmic) critical levels a and b* rejects $\omega=\omega_0$ if and only if $\Lambda_n>a$ for some n, $m\leqq n<N$, or $\Lambda_N>b$. Thus, letting

$$t=t_a=\inf\{n\geqq m: \Lambda_n>a\}, \tag{7.1}$$

the test takes $T=\min(N,t_a)$ observations and rejects $\omega=\omega_0$ if and only if either $t_a<N$ or $\Lambda_N>b$. The probability of a type I error is then

$$\alpha^*=P_{\omega_0}\{t_a<N\}+P_{\omega_0}\{t_a\geqq N, \Lambda_N>b\}, \tag{7.2}$$

which is typically much larger than α. Likelihood ratio and repeated likelihood ratio tests for $\omega\leqq\omega_0$ versus $\omega>\omega_0$ may be defined similarly. Approximations to the error probabilities of the repeated likelihood ratio tests are developed in the next section.

One may simplify the problem by supposing that $\omega_0=0$ and that $\psi'(0)=0=\psi(0)$; the simplification involves no loss of generality, since it may always be achieved by an appropriate reparameterization. Observe that then the dominating measure is $\lambda=G_0$. The closed convex support $\mathcal{Y}$ of G_0 is the convex hull of the support of G_0. Equivalently, $\mathcal{Y}$ is the smallest closed interval J for which $G_0\{J\}=1$. Observe that $\bar{Y}_n=(Y_1+\cdots+Y_n)/n\in\mathcal{Y}$ w.p. 1 for all $n\geqq 1$ and all $\omega\in\Omega$, for $\bar{Y}_n$ is a convex combination of elements of $\mathcal{Y}$, and the measures P_ω^n, $\omega\in\Omega$, are mutually absolutely continuous for $n\geqq 1$. Let $\underline{y}$ and $\bar{y}$ denote the endpoints of $\mathcal{Y}$, $-\infty\leqq\underline{y}<\bar{y}\leqq\infty$.

LEMMA 7.1. *The interior of $\mathcal{Y}$ is $\mathcal{Y}_0=\psi'(\Omega)$.*

Proof. If $\omega\in\Omega$, then $\bar{Y}_n\to\psi'(\omega)$ w.p. 1 (P_ω), so $\psi'(\omega)\in\mathcal{Y}$. Since $\psi'(\Omega)$ is open, $\psi'(\Omega)\subset\mathcal{Y}^0=(\underline{y},\bar{y})$. To complete the proof, it suffices to show that $\psi'(\omega)\to\underline{y}$ as $\omega\to\underline{\omega}$ and that $\psi'(\omega)\to\bar{y}$ as $\omega\to\bar{\omega}$. The assumption that Ω is open and Fatou's lemma combine to show that $\psi(\omega)\to\infty$ as $\omega\to\bar{\omega}$. It follows that, for any $z<\bar{y}$,

$$\int_{y\leqq z} yG_\omega\{dy\}\leqq \exp[-\psi(\omega)]\int_{y\leqq z} y e^{\omega y}G_0\{dy\}\to 0 \quad \text{as } \omega\to\bar{\omega}$$

and, consequently, $\liminf_{\omega\to\bar{\omega}} \psi'(\omega)\geqq z$. Since $z<\bar{y}$ was arbitrary, it now follows that $\psi'(\omega)\to\bar{y}$ as $\omega\to\bar{\omega}$, and the lower endpoint may be treated similarly.

In particular, it follows that $\phi(y)=J[0, \hat{\omega}(y)]$ is finite and twice continuously differentiable on the interior $\mathcal{Y}_0=(\underline{y}, \bar{y})$. Moreover, ϕ is convex on $\mathcal{Y}$, and $\phi(y)\geqq 0=\phi(0)$ for all y. It follows that ϕ is decreasing on $(\underline{y}, 0)$ and increasing on $(0, \bar{y})$ and it then follows that ϕ is continuous from above (below) at $\underline{y}(\bar{y})$, in the extended sense. Let

$$\underline{\phi}=\phi(\underline{y})\leqq\infty \quad\text{and}\quad \bar{\phi}=\phi(\bar{y})\leqq\infty.$$

LEMMA 7.2. *$\Lambda_n<\infty$ w.p. 1 (P_ω) for all $n\geqq 1$ and $\omega\in\Omega$.*

Proof. Λ_n can be infinite with positive probability only if $\bar{Y}_n=\bar{y}$ (resp. $\underline{y}$) with positive probability and $\bar{\phi}=\phi(\bar{y})=\infty$ (resp. $\underline{\phi}=\infty$). Except for an event of probability zero $\bar{Y}_n=\bar{y}$ (resp. $\underline{y}$) if and only if $Y_i=\bar{y}$ (resp. $\underline{y}$) for all $i=1,\cdots,n$, so Λ_n can be infinite with positive probability only if either $G\{\bar{y}\}>0$ and $\bar{\phi}=\infty$ or $G\{\underline{y}\}>0$ and $\underline{\phi}=\infty$. If $G\{\bar{y}\}>0$, then $\psi(\omega)\geqq\omega\bar{y}+\log G\{\bar{y}\}$ for all $\omega\in\Omega$, in which case $\phi(\bar{y})\leqq\log(1/G\{\bar{y}\})<\infty$. The lower endpoint may be analyzed similarly.

The next lemma illustrates a calculation which is used repeatedly in the next section.

LEMMA 7.3. *If $0\leqq\varepsilon<\bar{\phi}$, then there is a unique $\omega_\varepsilon\geqq 0$ for which $J(0,\omega_\varepsilon)=\varepsilon$. If $\underline{\omega}<\omega<\omega_\varepsilon$, then*

$$P_\omega\{\Lambda_n\geqq n\varepsilon, \bar{Y}_n>0\}\leqq\exp[-nJ(\omega,\omega_\varepsilon)],\qquad n\geqq 1,$$

and

$$P_\omega\{\Lambda_n\geqq n\varepsilon, \bar{Y}_n>0\}\sim[1/(\omega_\varepsilon-\omega)\sqrt{2\pi n\psi''(\omega_\varepsilon)}]\exp[-nJ(\omega,\omega_\varepsilon)]$$

as $n\to\infty$, if G_0 is not an arithmetic distribution.

Proof. Since $J(0,s)=\phi[\psi'(s)]$ increases strictly from 0 at $s=0$ to $\bar{\phi}$ as $s\to\bar{\omega}$, the equation $J(0,s)=\varepsilon$ has a unique positive solution $s=\omega_\varepsilon$ and, for $y>0$, $\phi(y)\geqq\varepsilon$ if and only if $y\geqq\psi'(\omega_\varepsilon)$. Thus,

$$\begin{aligned}P_\omega\{\Lambda_n\geqq n\varepsilon, \bar{Y}_n>0\}&=P_\omega\{\bar{Y}_n\geqq\psi'(\omega_\varepsilon)\}\\&=\int_{\bar{Y}_n\geqq\psi'(\omega_\varepsilon)}\exp\{(\omega-\omega_\varepsilon)n\bar{Y}_n-n[\psi(\omega)-\psi(\omega_\varepsilon)]\}\,dP_{\omega_\varepsilon}\\&=\exp[-nJ(\omega,\omega_\varepsilon)]\cdot\mathcal{I}_a,\end{aligned}$$

where

$$\mathcal{I}_n=\int_{\bar{Y}_n\geqq\psi'(\omega_\varepsilon)}\exp\{(\omega-\omega_\varepsilon)n[\bar{Y}_n-\psi'(\omega_\varepsilon)]\,dP_{\omega_\varepsilon}.$$

Since $\omega<\omega_\varepsilon$, $\mathcal{I}_n\leqq 1$, $n\geqq 1$, and the first assertion of the lemma follows.

The second assertion uses the fact that $Z_n = \sqrt{n}(\bar{Y}_n - \psi'(\omega_\varepsilon))/\sqrt{\psi''(\omega_\varepsilon)}$ is asymptotically standard normal, under ω_ε. The details are supplied only for the case that condition S is satisfied. Then Z_n has a density f_n for which $f_n(z) \to \Phi'(z)$ uniformly in $-\infty < z < \infty$ as $n \to \infty$, where Φ denotes the standard normal distribution. Let $\delta = (\omega_\varepsilon - \omega)\sqrt{\psi''(\omega_\varepsilon)}$. Then

$$\sqrt{n}\,\mathcal{I}_n = \int_0^\infty \exp(-\delta x) f_n\left(\frac{x}{\sqrt{n}}\right) dx \to \Phi'(0) \int_0^\infty \exp(-\delta x)\, dx = \frac{1}{\delta\sqrt{2\pi}},$$

by the dominated convergence theorem.

Of course, inequalities and approximations to $P_\omega\{\Lambda_n \geqq n\varepsilon,\ \bar{Y}_n < 0\}$ may be obtained by applying Lemma 7.3 to $-Y_k$, $k \geqq 1$; when combined with Lemma 7.3, the latter yield inequalities and approximations to $P_\omega\{\Lambda_n > n\varepsilon\}$. The result assumes an especially simple form when $\omega = 0$, for then $J(0, \omega_\varepsilon) = \varepsilon$. For example, Lemma 7.3 asserts

$$P_0\{\Lambda_n \geqq n\varepsilon,\ \bar{Y}_n > 0\} \leqq \exp(-n\varepsilon), \qquad 0 < \varepsilon < \bar{\phi}, \quad n \geqq 1. \tag{7.3}$$

Remarks and references. Lemma 7.3 provides an example of a large deviations calculation. For a proof, see Bahadur and Rao (1960). For applications of large deviations to large sample theory in statistics, see Bahadur (1971).

7.2. Error probabilities. Repeated likelihood ratio tests of $\omega \leqq 0$ versus $\omega > 0$, with initial sample size $m \geqq 1$, maximum sample size $N > m$, and (logarithmic) critical levels a and b, $0 < b \leqq a < \infty$, reject $\omega \leqq 0$ if either $\Lambda_n > a$ and $\bar{Y}_n > 0$ for some $n \in [m, N)$ or $\Lambda_N > b$ and $\bar{Y}_N > 0$. Thus, with $\Lambda_n^+ = \Lambda_n \cdot I_{\{\bar{Y}_n > 0\}}$ and

$$t^+ = t_a^+ = \inf\{n \geqq m : \Lambda_n^+ > a\}, \tag{7.4}$$

the test takes $t_a^+ \wedge N = \min(t_a^+, N)$ observations and rejects $\omega \leqq 0$ if either $t_a^+ < N$ or $t_a^+ \geqq N$ and $\Lambda_N^+ > b$. The reader may wish to compare the one-sided test with its two-sided analogue, described following (7.1). The one-sided test is of the form considered in Theorem 3.5, and it follows from Theorem 3.5 that the power function

$$\beta_+(\omega) = P_\omega\{t_a^+ < N\} + P_\omega\{t_a^+ \geqq N, \Lambda_N^+ > b\}$$

is an increasing function of ω. Thus, the maximum probability of a type I error is $\alpha_+^* = \beta_+(0)$. Approximations to the power function $\beta_+(\omega)$ are developed in Theorems 7.1–7.3 below, under the assumption that $N = N(a) \to \infty$ and $b = b(a) \to \infty$ as $a \to \infty$ with N/a bounded away from 0 and ∞ and $a - b = O(1)$ as $a \to \infty$. In addition, some results require that $m = m(a) \to \infty$ as $a \to \infty$. Approximations to the power of the two-sided test (7.1) are given, too.

Recall that $\Lambda_n = n\phi(\bar{Y}_n)$, $n \geqq 1$, and that ϕ is twice continuously differentiable on $\mathcal{Y}_0 = \psi'(\Omega)$. Thus, for each fixed $\omega \in \Omega$, $\Lambda_n = S_n + \xi_n$, $n \geqq 1$, where $\xi_n = \xi_n(\omega)$, $n \geqq 1$ are slowly changing with respect to (P_ω) and

$$\begin{aligned} S_n = S_n(\omega) &= n\phi[\psi'(\omega)] + n\phi'[\psi'(\omega)](\bar{Y}_n - \psi'(\omega)) \\ &= nJ(0, \omega) + n\omega(\bar{Y}_n - \psi'(\omega)), \qquad n \geqq 1, \end{aligned}$$

as in Chapter 6. By Lemma 6.4, $S_1 = S_1(\omega)$ has a nonarithmetic distribution for almost every $\omega \in \Omega$. So, with $t = t_a$ defined by (7.1), the excess $R_a = \Lambda_t - a$ has the limiting distribution H_ω of (6.10) for a.e. $\omega \in \Omega$, by Theorem 4.1. Moreover, it is easily seen that $P_\omega\{t_a^+ = t_a\} \to 1$ as $a \to \infty$ for all $\omega > 0$, so $R_a^+ = \Lambda_{t^+} - a$ has the same limiting distribution, if any, as R_a when $\omega > 0$.

The above derivation tacitly assumed that m remains fixed as $a \to \infty$. The modifications necessary when $m = m(a) \to \infty$ with a are described next.

LEMMA 7.4. *Suppose that $m = m(a) \to \infty$ as $a \to \infty$ with $m/a \to 1/\delta$, where $0 < \delta \leqq \infty$. Then, for a.e. ω with $J(0, \omega) < \delta$, R_a has the limiting distribution H_ω under P_ω; for all ω with $J(0, \omega) > \delta$, $R_a \to \infty$ in P_ω-probability; and for all $\omega > 0$, $P_\omega\{R_a^+ = R_a\} \to 1$.*

Proof. Let $t' = t'_a = \inf\{n \geqq 1: \Lambda_n > a\}$ and let $R'_a = \Lambda_{t'} - a$ for $a \geqq 0$. Then R'_a has the limiting distribution H_ω for a.e. $\omega \in \Omega$, by Theorem 4.1. Observe that $t_a = t'_a$ on $\{t'_a \geqq m\}$ and that $t'_a/a \to 1/J(0, \omega)$ in P_ω-probability for all $\omega \in \Omega$. Thus, if $J(0, \omega) < \delta$, then $P_\omega\{t_a = t'_a\} \geqq P_\omega\{t'_a \geqq m\} \to 1$, since $m/a \to 1/\delta < 1/J(0, \omega)$. It follows that $P_\omega\{R_a = R'_a\} \geqq P_\omega\{t_a = t'_a\} \to 1$; so R_a and R'_a have the same limiting distribution, if any. Next, if $\delta < \infty$ and $J(0, \omega) > \delta$, then $(\Lambda_m - a)/m \to J(0, \omega) - \delta > 0$ in P_ω-probability, so that $P_\omega\{t_a = m\} = P_\omega\{\Lambda_m > a\} \to 1$ and $R_a/m \to J(0, \omega) - \delta > 0$ in P_ω-probability.

Finally, if $\omega > 0$, then $P_\omega\{t_a^+ = t_a\} \to 1$ as $a \to \infty$, so that R_a^+ and R_a have the same limiting distribution, if any.

THEOREM 7.1. *Let $0 < \delta_0 < \delta_1 < \bar{\phi} = \phi(\bar{y})$ and suppose that $m \sim a/\delta_1$ and $N \sim a/\delta_0$ as $a \to \infty$. Then*

$$P_0\{t_a^+ < N\} \sim K_1^+ \sqrt{a}\, e^{-a} \quad \text{as } a \to \infty, \tag{7.5}$$

where

$$K_1^+ = K_1^+(\delta_0, \delta_1) = \frac{1}{\sqrt{2\pi}} \int_{\delta_0 < J(0,\omega) < \delta_1, \psi'(\omega) > 0} \mathcal{H}(\omega, 1) \sqrt{\frac{\psi''(\omega)}{J(0, \omega)}}\, d\omega$$

and $\mathcal{H}(\omega, \cdot)$ denotes the Laplace transform of the asymptotic distribution H_ω.

Proof. It is first shown that attention may be restricted to a compact subset of $\mathcal{Y}_0$. Let $\delta_1 < \phi_0 < \bar{\phi} \leqq \infty$ and let $C = \{y \in \mathcal{Y}_0: y \geqq 0, \phi(y) \leqq \phi_0\}$. Then C is a compact subset of $\mathcal{Y}_0$; and, by Lemma 3,

$$P_0\{t_a < N, \bar{Y}_{t^+} \notin C\} \leqq \sum_{k=m}^{N} P_0\{\bar{Y}_k > 0, \Lambda_k > k\phi_0\} \leqq N \exp(-m\phi_0),$$

which is of smaller order of magnitude than $\sqrt{a}\, e^{-a}$ as $a \to \infty$. Thus, it suffices to show that $P_0\{t_a^+ < N, \bar{Y}_{t^+} \in C\}$ is given by the right side of (7.5).

The remainder of the proof is similar to that of Theorem 6.2. Let q be a positive continuous density on Ω and let Q be the mixture

$$Q = \int_\Omega P_\omega q(\omega)\, d\omega.$$

Then

$$\frac{dQ^n}{dP_0^n}=\frac{1}{\sqrt{n}}\exp(\Lambda_n)u_n(\bar{Y}_n), \qquad n\geqq 1,$$

where

$$u_n(y)\to\sqrt{\frac{2\pi}{\psi''[\hat\omega(y)]}}q[\hat\omega(y)] \quad \text{uniformly with respect to } y\in C$$

as $n\to\infty$; see Lemma 6.2. In particular, it follows that $u_n(y)$ is bounded away from zero for $y\in C$ and $n\geqq 1$. Now, writing $t=t_a^+$ and $U_a=u_t(\bar{Y}_t)$,

$$(7.6)\qquad \begin{aligned} P_0\{t<N, \bar{Y}_t\in C\} &= \int_{t<N,\bar{Y}_t\in C}\frac{dP_0^t}{dQ^t}dQ \\ &= \int_{t<N,\bar{Y}_t\in C}\sqrt{t}\exp(-\Lambda_t)\frac{1}{U_a}dQ=\sqrt{a}\,e^{-a}\mathcal{J}_a, \end{aligned}$$

where

$$\mathcal{J}_a=\int_\Omega\left[\int_{t<N,\bar{Y}_t\in C}\sqrt{a^{-1}t}\exp(-R_a^+)\frac{1}{U_a}dP_\omega\right]q(\omega)\,d\omega.$$

Observe that the integrand is bounded, since $t<N\sim a\delta_0^{-1}$, $R_a^+\geqq 0$, and U_a is bounded below when $\bar{Y}_t\in C$. Denote the inner integral in $\mathcal{J}_a$ by $f_a(\omega)$, $\omega\in\Omega$, $a\geqq 0$. Then the limit of f_a is easily determined. If $\omega<0$, then $f_a(\omega)\to 0$, since $P_\omega\{\bar{Y}_t\in C\}\to 0$. If $\omega>0$ and $0<J(0,\omega)<\delta_0$, then $f_a(\omega)\to 0$, since $t/a\to 1/J(0,\omega)>1/\delta_0\sim N/a$ as $a\to\infty$; and, similarly, if $\omega>0$ and $J(0,\omega)>\delta_1$, then $f_a(\omega)\to 0$, since $R_a^+\to\infty$ in P_ω-probability, by Lemma 7.4. Finally, if $\omega>0$ and $\delta_0<J(0,\omega)<\delta_1$, then $t_a/a\to 1/J(0,\omega)$ and $U_a\to\sqrt{2\pi/\psi''(\omega)}\,q(\omega)$ in P_ω-probability; and, for a.e. such ω, $E_\omega[\exp(-R_a^+)]\to\mathcal{H}(\omega,1)$. Thus, since $J(0,\omega)=\delta_0$ and δ_1 for at most two ω each,

$$f_a(\omega)\to\frac{1}{\sqrt{2\pi}}\sqrt{\frac{\psi''(\omega)}{J(0,\omega)}}\frac{1}{q(\omega)}\mathcal{H}(\omega,1)I_{\{\delta_0<J(0,\omega)<\delta_1,\psi'(\omega)>0\}}$$

for a.e. $\omega\in\Omega$ as $a\to\infty$, and it follows easily from the dominated convergence theorem that $\mathcal{J}_a=\int_\Omega f_a q\,d\omega\to K_1^+=K_1^+(\delta_0,\delta_1)$.

There is a corresponding approximation for the two-sided test.

THEOREM 7.1′. *Let* $0<\delta_0<\delta_1<\min(\phi,\bar\phi)$ *and suppose that* $m\sim a\delta_1^{-1}$ *and* $N\sim a\delta_0^{-1}$ *as* $a\to\infty$. *With* $t=t_a$ *defined by* (7.1)

$$(7.7)\qquad P_0\{t_a<N\}\sim K_1\sqrt{a}\,e^{-a}\quad \text{as } a\to\infty,$$

where

$$(7.8)\qquad K_1=K_1(\delta_0,\delta_1)=\frac{1}{\sqrt{2\pi}}\int_{\delta_0<J(0,\omega)<\delta_1}\mathcal{H}(\omega,1)\sqrt{\frac{\psi''(\omega)}{J(0,\omega)}}\,d\omega.$$

The proof of Theorem 7.1′ is similar to that of Theorem 7.1 and will be omitted. In fact, replacing $\overset{+}{t}_a$ by t_a in the proof of Theorem 7.1 shows that $P_0\{t_a \leqq N, \bar{Y}_{t_a} > 0\} \sim K_1^+ \sqrt{a}\, e^{-a}$ as $a \to \infty$; Theorem 7.1′ then follows by applying this observation to $\pm Y_k$, $k \geqq 0$.

If $a = b$, then Theorems 7.1 and 7.1′ provide simple approximations to the significance levels $\alpha_+^* = P_0\{\overset{+}{t}_a < N+1\}$ and α^*. The accuracy of these approximations is investigated in § 7.4 in the normal case. Observe also that Theorem 7.1′ provides an approximate upper bound for $P_0\{\Lambda_s > a\}$ for all stopping times s for which $m \leqq s < N$.

If $b < a$, there is an additional term to be considered, namely,

$$P_0\{\overset{+}{t}_a \geqq N, \overset{+}{\Lambda}_N > b\} = P_0\{\overset{+}{\Lambda}_N > b\} - P_0\{\overset{+}{t}_a < N, \overset{+}{\Lambda}_N > b\}. \tag{7.9}$$

The probability that $\overset{+}{\Lambda}_N > b$ may be computed exactly in some cases—for example, the normal case. In others, it may be approximated from Lemma 7.3. The probability that $\Lambda_N > b$ and $\overset{+}{t}_a < N$ presents more challenge. Let

$$u_a(n, y) = P_\omega\{\overset{+}{t}_a \geqq n \mid \bar{Y}_n = y\}$$

be a version of the conditional probability which is independent of $\omega \in \Omega$. Thus, if B is any Borel subset of $\mathcal{Y}$, then

$$P_\omega\{\overset{+}{t}_a < N, \bar{Y}_N \in B\} = \int_{\bar{Y}_N \in B} [1 - u_a(N, \bar{Y}_N)]\, dP_\omega. \tag{7.10}$$

The probability of interest is of this form with $B = \{y: y > 0, N\phi(y) > b\}$.

If either condition A or condition S is satisfied then the conditional probabilities $u_a(n, y)$ may be defined as in § 5.1 (for sufficiently large n in the smooth case). Observe that $\overset{+}{t}_a$ is of the form considered in § 5.1, since $\overset{+}{t}_a = \inf\{n \geqq m: n\Delta(\bar{Y}_n) > a\}$, with $\Delta(y) = 0$ for $y \leqq 0$ and $\Delta(y) = \phi(y)$ for $y > 0$. Let

$$v(\omega, r) = P_\omega\{S_k(\omega) \geqq r, \text{ for all } k \geqq 1\}, \qquad -\infty < r < \infty, \quad \omega \in \Omega,$$

and let $\mathscr{C}(\omega)$ denote the continuity set of $v(\omega, \cdot)$ for $\omega \in \Omega$. Then Theorem 5.1 asserts: if $\omega > 0$, if $y = y_a \to \psi'(\omega)$, and if $n = n_a \to \infty$ as $a \to \infty$ with $y - \psi'(\omega) = O(1/\sqrt{n})$ and $n\phi(y) - a \to r \in \mathscr{C}(\omega)$, then $u_a(n, y) \to v(\omega, r)$.

In Theorems 7.3 and 7.4 below, it is assumed that m remains fixed as $a \to \infty$, although extensions are straightforward. It is also assumed that $N = N(a) = a\delta_0^{-1}$, where $0 < \delta_0 < \bar{\phi}$. Let ω_0 be the unique solution to

$$\omega_0 > 0 \quad \text{and} \quad \phi[\psi'(\omega_0)] = \delta_0. \tag{7.11}$$

THEOREM 7.2. *Let* $0 < \delta_0 < \bar{\phi}$ *and let* δ_0 *solve* (7.11). *Let* $a \to \infty$ *through integer multiples of* δ_0, *let* $N = N(a) = a\delta_0^{-1}$ *and let* $b = b(a) \to \infty$ *with* $0 < a - b \to c$, $0 < c < \infty$. *If condition* S *is satisfied, then*

$$P_0\{\overset{+}{t}_a < N, \overset{+}{\Lambda}_N > b\} \sim K_2^+ \frac{1}{\sqrt{N}} e^{-a} \quad \text{as } a \to \infty, \tag{7.12}$$

where

$$K_2^+ = K_2^+(\delta_0, c) = \frac{1}{\omega_0\sqrt{2\pi\psi''(\omega_0)}} \int_{-c}^{\infty} [1 - v(\omega_0, z)]\, e^{-z}\, dz.$$

Proof. The likelihood ratio $dP_0^N/dP_{\omega_0}^N$ may be written

$$\frac{dP_0^N}{dP_{\omega_0}^N} = \exp[-\omega_0 N\bar{Y}_N + N\psi(\omega_0)] = \exp[-NJ(0, \omega_0)]\exp(-Z_N),$$

where

$$Z_N = N\omega_0[\bar{Y}_N - \psi'(\omega_0)].$$

Thus,

$$P_0\{t_a^+ < N, \Lambda_N^+ > b\} = e^{-a} \int_{\Lambda_N^+ > b} [1 - u_a(N, \bar{Y}_N)] \exp(-Z_N)\, dP_{\omega_0} \tag{7.13}$$

by (7.10) and (7.11), since $NJ(0, \omega_0) = N\phi[\psi'(\omega_0)] = a$. Denote the integral in (7.13) by $\mathcal{I}_a$, $a > 0$. Let F_N denote the distribution of Z_N when $\omega = \omega_0$, let

$$y_N(z) = \psi'(\omega_0) + \frac{1}{N\omega_0} z \quad \text{and} \quad \phi_N(z) = N\{\phi[y_N(z)] - \phi[\psi'(\omega_0)]\}$$

for $-\infty < z < \infty$ and let $D = D_a = \{z : 0 < y_N(z) \in \mathcal{Y}$ and $\phi_N(z) > b - a\}$. Then, since $N\phi[\psi'(\omega_0)] = a$, one has

$$\mathcal{I}_a = \int_{D_a} [1 - u_a(N, y_N(z))]\, e^{-z} F_N\{dz\}, \qquad a > 0.$$

Now, $\phi_N(z)$ is increasing for $y_N(z) > 0$ and $\phi_N(z) \to z$ uniformly in z on compact subintervals of $(-\infty, \infty)$, so D_a is an interval of the form $(-c_a, \infty)$, where $c_a \to c$ as $a \to \infty$. Moreover, $y_N(z) - \psi'(\omega_0) = O(1/N) = O(1/\sqrt{N})$ and $N\phi[y_N(z)] - a = \phi_N(z) \to z$ as $a \to \infty$, so $u_a[N, y_N(z)] \to v(\omega_0, z)$ for all z. Finally, F_N as a continuous density f_N for sufficiently large N, and $\sqrt{N} f_N(z) \to 1/\sigma_0\sqrt{2\pi}$ as $N \to \infty$ with $\sigma_0 = \omega_0\sqrt{\psi''(\omega_0)}$. Since $c < \infty$, it now follows from the dominated convergence theorem that

$$\sqrt{N}\mathcal{I}_a \to \frac{1}{\sigma_0\sqrt{2\pi}} \int_{-c}^{\infty} [1 - v(\omega_0, z)]\, e^{-z}\, dz \quad \text{as } a \to \infty,$$

as asserted in (7.12).

A similar result holds for testing $\omega = 0$.

THEOREM 7.2′. *Under the hypotheses of Theorem* 7.2, (7.12) *holds with* t_a^+ *replaced by* t_a.

As corollaries to Theorems 7.2 and 7.2′, one may complete the approximations to the significance levels α_+^* and α^* in the case $b < a$. For simplicity, the result is stated only for the one-sided hypothesis, $\omega \leqq 0$. Let $K_0^+ = 1/\omega_0\sqrt{2\pi\psi''(\omega_0)}$.

Then, $P_0\{\Lambda_N^+ > b\} \sim K_0^+ (1/\sqrt{N})\, e^{-b}$ as $a \to \infty$, by Lemma 7.3, so

$$P_0\{t_a^+ \geqq N, \Lambda_N^+ > b\} \sim K_0^+ \frac{1}{\sqrt{N}} e^{-b} - K_2^+ \frac{1}{\sqrt{N}} e^{-a}$$

and

$$\alpha_+^* \approx K_1^+ \sqrt{a}\, e^{-a} + K_0^+ \frac{1}{\sqrt{N}} e^{-b} - K_2^+ \frac{1}{\sqrt{N}} e^{-a} \quad \text{as } a \to \infty, \tag{7.14}$$

by (7.9) and Theorems 7.1 and 7.2. Observe that the second two terms on the right side of (7.14) are of smaller order of magnitude than the first as $a \to \infty$; so, in a mathematical sense, they could be neglected. It is recommended that they be retained in numerical calculations, however. It is also recommended that δ_0 and δ_1 be defined as $\delta_0 = a/N$ and $\delta_1 = a/m$ in the calculation of K_0^+, K_1^+ and K_2^+.

It is easily seen that $\beta_+(\omega) \to 0$ for all $\omega < \omega_0$ and $\beta_+(\omega) \to 1$ for all $\omega > \omega_0$, under the limiting operation described in Theorem 7.2. The next theorem determines the rate of convergence. Observe that

$$\begin{aligned} \beta_+(\omega) &= P_\omega\{\Lambda_N^+ > b\} + P_\omega\{t_a^+ < N, \Lambda_N^+ \leqq b\}, \\ 1 - \beta_+(\omega) &= P_\omega\{\Lambda_N^+ \leqq b\} - P_\omega\{t_a^+ < N, \Lambda_N^+ \leqq b\} \end{aligned} \tag{7.15}$$

for all $\omega \in \Omega$. As above, the distribution of Λ_N^+ may be determined exactly in some cases, and approximated from Lemma 7.3 in others. Thus, interest centers on $P_\omega\{t_a^+ < N, \Lambda_N^+ \leqq b\}$ with $b \leqq a$.

THEOREM 7.3. *Suppose that condition* S *is satisfied. Let* $0 < \delta_0 < \bar{\phi}$, *define* ω_0 *by* (7.11), *let* $N = N(a) = a\delta_0^{-1}$, $a > 0$, *let* $a \to \infty$ *through integer multiples of* δ_0 *and let* $b = b(a) \to \infty$ *with* $a - b \to c$, $0 \leqq c < \infty$. *If* $\omega > 0$ *and* $\omega \neq \omega_0$, *then*

$$P_\omega\{t_a^+ < N, \Lambda_N \leqq b\} \sim K_3^+ \frac{1}{\sqrt{N}} \exp[-NJ(\omega, \omega_0)] \quad \text{as } a \to \infty, \tag{7.16}$$

where

$$K_3^+ = K_3^+(c, \omega, \delta_0) = \frac{1}{\omega_0 \sqrt{2\pi\psi''(\omega_0)}} \int_{-\infty}^{-c} [1 - v(\omega_0, z)] \exp\left[\left(\frac{\omega - \omega_0}{\omega_0}\right) z\right] dz.$$

Proof. The proof is similar to that of Theorem 7.2. First, one finds from (7.10)

$$P_\omega\{t_a^+ < N, \Lambda_N^+ \geqq b\} = \exp[-NJ(\omega, \omega_0)] \mathcal{I}_a,$$

where

$$\begin{aligned} \mathcal{I}_a &= \int_{\Lambda_N^+ \leqq b} [1 - u_a(N, \bar{Y}_N)] \exp\left[\left(\frac{\omega - \omega_0}{\omega_0}\right) Z_N\right] dP_{\omega_0} \\ &= \int_{D_a} [1 - u_a(N, y_N(z))] \exp\left[\left(\frac{\omega - \omega_0}{\omega_0}\right) z\right] F_N\{dz\}, \end{aligned}$$

with $y_N(z)$, $\phi_N(z)$ and F_N as in the proof of Theorem 7.2 and $D_a=\{z: y_N(z)\leqq 0$ or $\phi_N(z)\leqq b\}$. In this case D_a is an interval with left endpoint $-\infty$ and right endpoint $c_a\to -c$. If $\omega>\omega_0$, then it follows directly from the dominated convergence theorem that $\sqrt{N}\mathcal{I}_a\to K_3^+$, as in the proof of Theorem 7.2. If $0<\omega<\omega_0$, then additional justification is needed. The details are omitted.

Again, a similar result holds for testing $\omega=0$.

THEOREM 7.3′. *Under the hypotheses of Theorem* 7.3, (7.16) *holds with* t_a^+ *replaced by* t_a.

The constants K_1^+ and K_1 may be computed by numerical integration, using the results of §§ 2.4 and 3.1 to compute the Laplace transform $\mathcal{H}(\omega, 1)$. The calculation of K_2^+ and K_3^+ presents additional difficulties. Let $S_k=S_k(\omega_0)=kJ(0,\omega_0)+k\omega_0(\bar{Y}_k-\psi'(\omega_0))$, $k\geqq 1$, and let $M_0=\min\{S_0, S_1,\cdots\}$. Then

$$1-v(\omega_0, z)=P_{\omega_0}\{M_0<z\},\qquad -\infty<z<0.$$

Thus, $1-v(\omega_0, z)$ may be approximated from Theorem 3.1 with $a=\infty$ and $b=-z$ as

$$1-v(\omega_0, z)\sim \mathcal{H}_0^-(1)\,e^z \quad \text{as } z\to-\infty, \tag{7.17}$$

where $\mathcal{H}_0^-$ denotes the Laplace transform of the asymptotic distribution of residual waiting time for the random walk $-S_k$, $k\geqq 0$, when $\omega=0$. See § 3.1. Moreover, by Theorem 3.3,

$$\mathcal{H}_0^-(1)=\frac{J(0,\omega_0)}{J(\omega_0, 0)}\mathcal{H}(\omega_0, 1)=\frac{\delta_0\mathcal{H}(\omega_0,1)}{\psi(\omega_0)}, \tag{7.18}$$

where $\mathcal{H}(\omega_0,\cdot)$ is the asymptotic Laplace transform of residual waiting time for S_k, $k\geqq 0$, when $\omega=\omega_0$. Substituting (7.17) and (7.18) into the definition of K_3^+ now yields

$$K_3^+(\omega,\delta_0, c)\sim\frac{1}{\omega\sqrt{2\pi\psi''(\omega_0)}}\,\frac{\delta_0\mathcal{H}(\omega_0,1)}{\psi(\omega_0)}\exp\left(-\frac{\omega c}{\omega_0}\right) \tag{7.19}$$

as $c\to\infty$. We use (7.19) for moderate values of c, too.

In the special case that $c=0$, K_3^+ may be computed as an exponential series. Indeed, a simple integration by parts shows that

$$K_3^+(\omega,\delta_0,0)=\left(\frac{1}{\omega-\omega_0}\right)\sqrt{\frac{1}{2\pi\psi''(\omega_0)}}E_{\omega_0}\left\{1-\exp\left[\left(\frac{\omega-\omega_0}{\omega_0}\right)M_0\right]\right\},$$

$$E(e^{\gamma M_0})=\exp\left\{\sum_{k=1}^{\infty}\frac{1}{k}\int_{S_k\leqq 0}[e^{\gamma S_k}-1]\,dP_{\omega_0}\right\},\qquad \gamma>-1,$$

by Spitzer's formula.

Similar considerations apply to K_2^+. One finds

$$\int_0^\infty [1-v(\omega_0, r)]\,e^{-r}\,dr = 1-\delta_0\mathcal{H}(\omega_0, 1),$$

$$\int_{-c}^0 [1-v(\omega_0, r)]\,e^{-r}\,dr \sim \frac{\delta_0\mathcal{H}(\omega_0, 1)}{\psi(\omega_0)}c \quad \text{as } c\to\infty,$$

$$K_2^+(\delta_0, c) \sim \frac{1}{\omega_0\sqrt{2\pi\psi''(\omega_0)}}\left\{\frac{\delta_0\mathcal{H}(\omega_0, 1)}{\psi(\omega_0)}c+1-\delta_0\mathcal{H}(\omega_0, 1)\right\}. \tag{7.20}$$

Example 7.2. Suppose that G_ω, is the exponential distribution with failure rate $|\omega|$, $-\infty<\omega<0$, and consider testing the null hypotheses $|\omega|\leqq 1$ versus $|\omega|>1$. Then the reparameterization

$$\omega' = -(1+\omega) \quad \text{and} \quad Y' = 1-Y$$

brings the problem into the form considered in Theorems 7.1–7.3. For brevity, the results are stated directly in terms of the original parameterization. Let

$$\psi(\omega) = \log\frac{1}{|\omega|}-(1-|\omega|), \qquad -\infty<\omega<0.$$

Then

$$J(-1, \omega) = (1+\omega)\frac{1}{|\omega|}+\log|\omega|$$

and

$$\mathcal{H}(\omega, 1) = \frac{(|\omega|-1)-\log|\omega|}{|\omega|\log|\omega|-(|\omega|-1)}$$

for $\omega<-1$; see Example 3.2. For a repeated significance test with sample sizes $1\leqq m<N$ and critical levels $0<b\leqq a$, let $c=a-b$, $\delta_0=a/N$ and $\delta_1=a/m$. Then Theorem 7.1 is applicable with

$$K_1^+ = \int_{\delta_0<J(-1,\omega)<\delta_1,\,\omega<-1} \mathcal{H}(\omega, 1)\sqrt{\frac{\psi''(\omega)}{J(-1, \omega)}}\,d\omega.$$

Next, let $\omega_0<-1$ solve $J(-1, \omega_0)=\delta_0$. Then $K_2^+(\delta_0, c)$ and $K_3^+(\omega, \delta_0, c)$ may be computed directly from (7.19) and (7.20). In this example, the approximate equalities of (7.19) and (7.20) are, in fact, equalities, since S_1 has an exponential left tail.

Remarks and references. Theorem 7.1 was established by Woodroofe (1976b) in the normal case, using a different proof. The proof given here follows that used by Lai and Siegmund (1977), again in the normal case. Theorems 7.2 and 7.3 were established by Siegmund (1977), (1978) in the normal case.

7.3. The expected sample size. In this section approximations to the expected value of the sample size $T_a=\min(t_a, N)$ are developed. For simplicity, the null hypothesis is taken to be $\omega=0$ and m is taken to be 1 throughout most of the development. Let

$$\mu_\omega = J(0,\omega), \qquad \rho_\omega = \int_0^\infty rH_\omega\{dr\}, \qquad \omega\in\Omega,$$

where H_ω is as in (6.10).

THEOREM 7.4. *Define t_a by (7.1). If $S_1(\omega)$ does not have an arithmetic distribution, then*

$$E_\omega(t_a)=\frac{1}{\mu_\omega}\left[a+\rho_\omega-\frac{1}{2}\right]+o(1) \quad \text{as } a\to\infty. \tag{7.21}$$

Proof. Theorem 7.4 may be deduced from Theorem 4.5 by verifying the conditions (4.10)–(4.16). The details of the verification are similar to, but simpler than, those presented in the proof of Theorem 6.3 and have been omitted. The verification of (4.16) is effectively included in the proof of Theorem 7.5 below, however.

If $\omega\neq 0$ and $S_1(\omega)$ does have an arithmetic distribution, then (7.21) holds as $a\to\infty$ through multiples of the span $d(\omega)$, with a different value for ρ_ω (cf. Theorem 6.3).

THEOREM 7.5. *Suppose that $N=N_a\sim a/\delta_0$ as $a\to\infty$, where $0<\delta_0<\infty$. Then*

$$E_\omega(T_a)=\begin{cases} N+o(1), & J(0,\omega)<\delta_0 \\ E_\omega(t_a)+o(1), & J(0,\omega)>\delta_0\end{cases} \quad \text{as } a\to\infty. \tag{7.22}$$

Proof. If $J(0,\omega)<\delta_0$, then $0\leqq N-E_\omega(T_a)\leqq NP_\omega\{t_a\leqq N\}$, so it suffices to show that $P_\omega\{t_a\leqq N\}=o(1/N)$ as $a\to\infty$. Let $2\varepsilon=\delta_0-J(0,\omega)$ and let B_a be the event that $|\bar{Y}_N-\psi'(\omega)|\leqq\varepsilon/(1+|\omega|)$. Then $P_\omega(B_a')=O(1/N)$, so it suffices to show that $P_\omega\{t_a\leqq N, B_a\}=o(1/N)$. Now,

$$P_\omega\{t_a\leqq N, B_a\}=\int_{t_a\leqq N, B_a}\frac{dP_\omega^N}{dP_0^N}\,dP_0\leqq \exp\{N[J(0,\omega)+\varepsilon]\}P_0\{t_a\leqq N\}$$

as in the proof of Theorem 6.3, and $P_\omega\{t_a\leqq N\}\leqq Ne^{-a}$ by (7.3). This establishes (7.22) in the special case that $J(0,\omega)<\delta_0$.

When $J(0,\omega)>\delta_0$, the argument is slightly simpler. Then

$$0\leqq E_\omega(t_a)-E_\omega(T_a)=\int_{t_a>N}(t_a-N)\,dP_\omega\leqq\int_{t_a>2N}(t_a-N)\,dP_\omega+NP_\omega\{\Lambda_N\leqq a\},$$

which tends to zero as $a\to\infty$, by (4.8) and Lemma 7.3.

By exercising more care in the case $\omega=0$, one may show

$$N-E_0(T_a)=\int_{t\leqq N}(N-t)\,dP_0\sim Ka^{3/2}e^{-a} \quad \text{as } a\to\infty, \tag{7.23}$$

where

$$K=\frac{1}{\sqrt{2\pi}}\int_{J(0,\omega)>\delta_0}\left[\frac{1}{\delta_0}-\frac{1}{J(0,\omega)}\right]\mathscr{H}(\omega,1)\sqrt{\frac{\psi''(\omega)}{J(0,\omega)}}\,d\omega.$$

The reader may supply the details, which are similar to the proof of Theorem 7.1.

Numerical evidence, presented in the next section, suggests that the approximation (7.22) may be adequate when ω is near 0 or $J(0, \omega)$ is much larger than δ_0, but the approximation is not uniform near the solution to the equation $J(0, \omega) = \delta_0$ and may substantially overestimate $E_\omega(T_a)$ there. The next theorem indicates the nature of the nonuniformity.

THEOREM 7.6. *Let $0 < \delta_0 < \bar{\phi}$ and let $\omega_0 > 0$ solve $J(0, \omega_0) = \delta_0$. Further, let $N = [a/\delta_0]$. Then*

$$E_{\omega_0}(T_a) = N - \frac{\omega_0}{\delta_0}\sqrt{\frac{\psi''(\omega_0)}{2\pi}N} + o(\sqrt{N}) \quad \text{as } a \to \infty. \tag{7.24}$$

Proof. Let $t_a^* = (t_a - N)/\sqrt{N}$, so that t_a^* is asymptotically normal with mean 0 and variance $\omega_0^2\psi''(\omega_0)/\delta_0^2$, under P_{ω_0}. Then

$$T_a = N - t_a^{*-}\sqrt{N},$$

where $^-$ denotes negative part, so it suffices to show that t_a^{*-}, $a \geqq 1$, are uniformly integrable. Now

$$\int_{t_a \leqq N/2} |t_a^*| \, dP_{\omega_0} \leqq \sqrt{N} P_{\omega_0}\left\{t_a \leqq \frac{N}{2}\right\} \to 0 \quad \text{as } a \to \infty,$$

and, on $\{N/2 \leqq t_a \leqq N\}$,

$$\frac{1}{\sqrt{N}}\left(t_a - \frac{a}{\delta_0}\right) = \frac{-1}{\delta_0\sqrt{N}}(S_{t_a} - \delta_0 t_a) + \frac{1}{\delta_0\sqrt{N}}(R_a - \xi_{t_a})$$

$$\geqq \frac{-1}{\delta_0\sqrt{N}}|S_{t_a} - \delta_0 t_a| - \max_{N/2 \leqq k \leqq N} \frac{\xi_k}{\delta_0\sqrt{N}},$$

which is uniformly integrable under P_{ω_0}.

It may be possible to refine (7.24).

CONJECTURE. *Let $0 < \delta_0 < \bar{\phi}$, let $\omega_0 > 0$ solve $J(0, \omega_0) = \delta_0$ and let $N = a/\delta_0$. If $S_1(\omega_0)$ does not have an arithmetic distribution, then*

$$E_{\omega_0}(T_a) = \frac{1}{\delta_0}\left[a - \omega_0\sqrt{\frac{\psi''(\omega_0)}{2\pi}N}\right] + O(1) \tag{7.25}$$

as $a \to \infty$ through integer multiples of δ_0.

Outline of the proof. In this case, write

$$E_{\omega_0}(t_a) - E_{\omega_0}(T_a) = \int_{t_a > N} E(t_a - N | \mathscr{A}_N) \, dP_{\omega_0}, \tag{7.26}$$

where $\mathscr{A}_N = \sigma\{Y_1, \cdots, Y_N\}$. Let $b = a - Z_N$. Then $b > 0$ on $\{t_a > N\}$ and

$$t_a - N = \inf\{n \geqq 1: Z_{N+n} - Z_N > b\},$$

so

$$E_{\omega_0}(t_a - N|\mathcal{A}_N) = \frac{1}{\mu_{\omega_0}}[b + \rho_{\omega_0}] + o(1)$$

is suggested. When this relation is substituted into (7.26) and integrated, (7.25) may be obtained.

Remarks and references. Relation (7.23) is adapted from Siegmund (1977). Theorem 7.6 is adapted from Woodroofe (1976b). It would be desirable to determine the $O(1)$ term in (7.25) and to have an expansion for $E_\omega(t_a)$ which is uniform in a neighborhood of ω_0.

See Lorden (1973) for bounds on the expected sample size of a closely related procedure.

7.4. The normal case. In this section, the results of §§ 7.2 and 7.3 are specialized to the normal case. Thus, suppose that G_ω is the normal distribution with unknown mean ω, $-\infty < \omega < \infty$, and unit variance, and that the hypotheses $\omega = 0$ versus $\omega \neq 0$ are of interest. Then

$$\psi(\omega) = \frac{\omega^2}{2} \quad \text{and} \quad J(\omega_1, \omega_2) = \frac{(\omega_2 - \omega_1)^2}{2}$$

for $-\infty < \omega$, ω_1, $\omega_2 < \infty$. The random walks $S_k(\omega) = dP_\omega^k / dP_0^k$ are $S_k = \frac{1}{2}\omega^2 k + \omega k(\bar{Y}_k - \omega)$, $k \geqq 1$. The Laplace transform of the asymptotic distribution of residual waiting time $\mathcal{H}(\omega, 1)$ was computed in Example 3.1 as $\mathcal{H}(\omega, 1) = 2\omega^{-2}\kappa(\omega^2/2)$ for $\omega \neq 0$, where

$$\kappa(\varepsilon) = \exp\left\{-2 \sum_{k=1}^{\infty} \frac{1}{k} \Phi\left[-\sqrt{\frac{\varepsilon k}{2}}\right]\right\}, \qquad \varepsilon > 0,$$

and a brief table of $\mathcal{H}(\omega, 1)$ was also included in Example 3.1.

In the normal case, Theorems 7.1 and 7.1′ remain valid when m remains fixed as $a \to \infty$. The constant K_1^+ of Theorem 7.1 may be written

$$K_1^+(\delta_0, \delta_1) = \frac{1}{2\sqrt{\pi}} \int_{\delta_0}^{\delta_1} \kappa(\varepsilon)\varepsilon^{-2}\, d\varepsilon$$

for $0 < \delta_0 < \delta_1 < \infty$, as is easily seen from the change of variables $\varepsilon = J(0, \omega) = \omega^2/2$. Of course, $K_1 = 2K_1^+$ in Theorem 7.1′, by symmetry. Observe that $K_1^+(\delta_0, \delta_1) = K_1^+(1, \delta_1) + K_1^+(\delta_0, 1)$. Selected values of $K(1, \delta)$ are listed in Table 7.1 below. In applications, m, N and a may be given, in which case δ_0 and δ_1 are determined.

The constants K_2^+ and K_3^+ may be simply approximated, too. If $0 < \delta_0 < \infty$, then $\omega_0 = \sqrt{2\delta_0}$ solves $J(0, \omega_0) = \delta_0$, and

$$K_2^+(\delta_0, c) \approx \frac{1}{\omega_0\sqrt{2\pi}}[\mathcal{H}(\omega_0, 1)(c - \delta_0) + 1],$$

$$K_3^+(\delta_0, \omega, c) \approx \frac{1}{\omega\sqrt{2\pi}} \mathcal{H}(\omega_0, 1) \exp\left(-\frac{\omega c}{\omega_0}\right),$$

at least for large c. Moreover, when $c=0$ and $\omega > \omega_0$, then K_3^+ may be computed explicitly as

$$K_3^+(\delta_0, \omega, 0) = \frac{1}{(\omega-\omega_0)\sqrt{2\pi}}\left[1 - C\left(\frac{\omega-\omega_0}{\omega_0}\right)\right],$$

where

$$C(\gamma) = \exp\left(\sum_{k=1}^{\infty} \frac{1}{k}\left\{m_0(\gamma)^k \Phi\left[-\sqrt{\frac{\delta_0 k}{2}}(1+2\gamma)\right] - \Phi\left[-\sqrt{\frac{\delta_0 k}{2}}\right]\right\}\right), \qquad \gamma > -1,$$

and $m_0(\gamma) = \exp\{\delta_0\gamma + \delta_0\gamma^2\}$. Selected values of $K_3^+(\delta_0, \omega, 0)$ are given in Table 7.2.

The computation of the mean $\rho(\omega)$ was discussed in Example 3.1, and a brief table was included there.

To assess the accuracy of the approximations, one may compare the approximations with the "exact" values, computed by repeated numerical integrations by Armitage, McPherson and Rowe (1969) and McPherson and Armitage (1971).

TABLE 7.1
Values of $K_1(\delta, 1)$

δ	$K_1(\delta, 1)$	δ	$K_1(1, \delta)$
.01	1.9230	1.5	.0930
.02	1.5684	2.0	.1491
.03	1.3673	2.5	.1869
.04	1.2280	3.0	.2142
.05	1.1222	3.5	.2349
.06	1.0374	4.0	.2512
.07	.9668	4.5	.2643
.08	.9067	5.0	.2750
.09	.8544	6	.2916
.10	.8092	7	.3038
.11	.7670	8	.3130
.12	.7299	9	.3203
.13	.6961	10	.3261
.14	.6652	11	.3308
.15	.6368	12	.3348
.20	.5215	13	.3381
.25	.4352	14	.3409
.30	.3693	15	.3433
.40	.2692	20	.3513
.50	.1964	∞	.4013
.60	.1403		
.70	.0953		
.80	.0582		
.90	.0229		

For $\varepsilon \geqq 20$, use $\kappa(\varepsilon) \approx 1$. The computations were done on the University of Michigan's Amdahl 470, using the trapezoid rule with subdivision of .0001.

TABLE 7.2
Values of $K_3^+(\delta_0, \omega, 0)$

ω \ ω_0	.25	.50	.75	1.00	1.25	1.50
.25	1.3756	1.1854	1.0214	.8804	.7591	.6550
.50	.6860	.5894	.5060	.4342	.3723	.3195
.75	.4563	.3910	.3346	.2860	.2443	.2084
1.00	.3415	.2919	.2491	.2123	.1806	.1534
1.25	.2727	.2327	.1981	.1682	.1426	.1207
1.50	.2269	.1932	.1641	.1391	.1175	.0991
1.75	.1943	.1651	.1400	.1183	.0998	.0838
2.00	.1698	.1441	.1219	.1029	.0865	.0725
2.50	.1355	.1147	.0968	.0814	.0682	.0569
3.00	.1128	.0953	.0802	.0673	.0562	.0467

Here $\omega_0 = \sqrt{2\delta_0}$. The computations were done on an Apple II microcomputer using formula (26.2.17) of Abramowitz and Stegun (1970) to compute the standard normal distribution function.

TABLE 7.3
The accuracy of the approximations when $a = b$

a	N	δ_0	ω	α^*	$1-\beta$	$E_\theta(T)$
4.351	200	.0218	.3	.049	.059	85.4
				.051	.062	89.6
4.176	111	.0376	.4	.049	.056	47.2
				.051	.058	49.1
4.033	71	.0568	.5	.050	.053	29.9
				.051	.055	30.9
3.920	49	.0800	.6	.049	.051	20.7
				.050	.053	21.2
3.726	28	.1331	.8	.049	.043	11.7
				.049	.045	11.7

The upper figure is "exact," from the numerical integrations of McPherson and Armitage (1971); the lower figure is approximate, obtained by specializing Theorems 7.1, 7.3 and 7.4.

Such a study was undertaken by Siegmund (1977). A portion of his results are reproduced in Table 7.3. It appears that the approximations are quite accurate, at least in the normal case.

To illustrate the use of the formulas and tables, we compare three tests of the hypothesis $\omega = 0$, each with approximate level .05 and approximate power .95 when $\omega = .5$. First, a fixed sample size test with sample size $N_0 = 52$ has these properties. Next, according to Table 7.3, a repeated significance test with $m = 1$, $N = 71$, and $a = b = 4.033$ has these properties, too. Finally, by trial and error with Tables 7.1 and 7.2, one finds that a repeated significance test with $m = 1$, $N = 60$, $a = 5$, and $b = 2.5$ has these properties, too. In fact,

$$P_0\{t < N\} \sim .025 \sim P_0\{t \geqq N, \Lambda_N > b\},$$

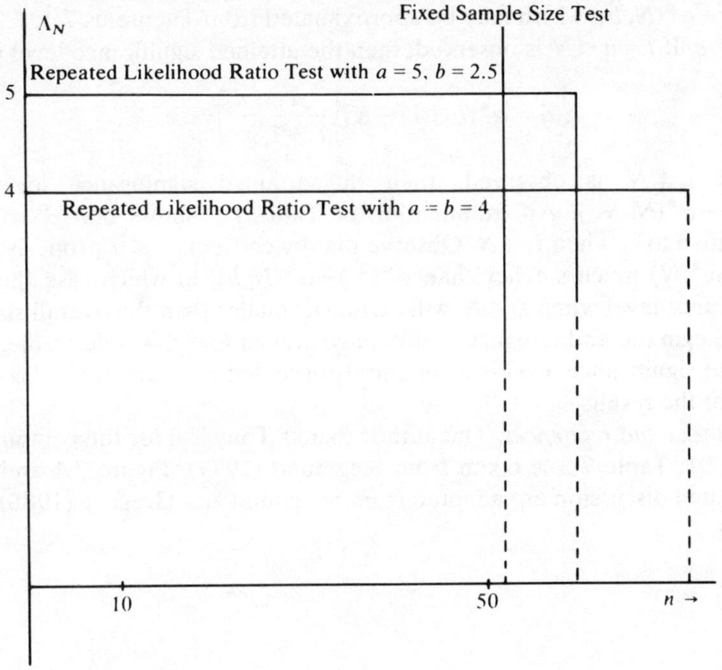

FIG. 7.1. *Three tests of* $\omega = 0$. - - - *Accept* $\omega = 0$; —— *Reject.*

so that the type I error is divided evenly between rejecting with $t<N$ and rejecting with $t \geqq N$. The relationships among these three tests are illustrated in Fig. 7.1.

The fixed sample size test has the smallest maximum sample size, but does not allow the possibility of stopping early when $|\omega|$ is large. The repeated significance test with $a = b$ does offer the possibility of stopping early when $|\omega|$ is large, but increases the maximum sample size by 40%. Moreover, if $|\omega|$ is small, then one is virtually certain to continue the test to its end. The repeated significance test with $0<b<a$ offers a compromise between the two, for it allows the possibility of stopping early when $|\omega|$ is large with a smaller increase in the maximum sample size. Of course, the expected sample size is larger for large $|\omega|$ when $b<a$, since a must be increased. The appropriate test must depend on the particular application—in particular, the extent to which stopping early is desirable when $|\omega|$ is large.

Theorems 7.1, 7.2 and 7.3 may be used to approximate the attained significance level of repeated significance tests, the smallest significance level at which one would reject the null hypothesis. Let

$$\alpha^*(n) = P_0\{t_a < n\} \quad \text{and} \quad \alpha^*(N, b) = P_0\{t_a \geqq N, \Lambda_N > b\}.$$

If repeated likelihood ratio tests with sample sizes $1 \leqq m < N$ and critical levels $0<b \leqq a$ are performed, then the overall significance level is $\alpha^* =$

$\alpha^*(N)+\alpha^*(N, b)$, which may be approximated from Theorems 7.1, 7.2 and 7.3, as above. If $t=n<N$ is observed, then the attained significance level is

$$\hat{\alpha}=\alpha^*(t_a+1) \sim K_1\left[\frac{a}{t_a+1}, \frac{a}{m}\right] \sqrt{a}\, e^{-a},$$

and if $t_a \geqq N$ is observed, then the attained significance level is $\alpha=\alpha^*(N)+\alpha^*(N, \Lambda_N)$. An examination of Table 7.1 shows that $\hat{\alpha}$ is relatively insensitive to t_a when $t_a<N$. Observe that by choosing $b<a$ properly, one may make $\alpha^*(N)$ much smaller than $\alpha^*(N)+\alpha^*(N, b)$, in which case the attained significance level when $t_a<N$ will be much smaller than the overall significance level. Siegmund and Gregory (1980) have argued that this is desirable, since the attained significance level is a commonly accepted measure of the "convincingness" of the results.

Remarks and references. The author thanks Tony Tai for the computations in Table 7.1. Table 7.3 is taken from Siegmund (1977). Figure 7.1 and much of the related discussion are adapted from Siegmund and Gregory (1980).

CHAPTER 8

Multiparameter Problems

8.1. Multiparameter exponential families. In this chapter some of the results of Chapter 7 are extended to a multiparameter context. To simplify complicated expressions, it is convenient to adopt some different notation. In this chapter, prime (′) denotes transpose, and R^p denotes the set of all $p \times 1$ matrices $x = (x_1, \cdots, x_p)'$. Thus, if $x, y \in R^p$, then the inner product between x and y is $x'y = x_1y_1 + \cdots + x_py_p$, and the norm of x is $\|x\| = \sqrt{x'x}$. Differentiation is denoted by D, and ∇ is used to denote the gradient.

If $\Omega \subset R^p$, then a family G_ω, $\omega \in \Omega$, of probability distributions over R^p is said to be an exponential family if and only if

$$G_\omega\{dy\} = \exp[\omega'y - \psi(\omega)]\lambda\{dy\}, \qquad y \in R^p, \quad \omega \in \Omega, \tag{8.1}$$

for some sigma-finite measure λ. Here the normalizing constant has been written in the form $\exp[\psi(\omega)] = \int \exp(\omega'y)\lambda\{dy\}$ and the natural parameter space is defined to be the set of all $\omega \in R^p$ for which the latter integral is finite. It may be shown that the natural parameter space is a convex subset of R^p, that ψ is convex on Ω, and that ψ is infinitely differentiable on the interior Ω^0 of Ω. See, for example, Lehmann (1959, § 2.7). Throughout this chapter, Ω denotes the natural parameter space of the family (8.1), and Ω is assumed to be open in R^p.

If $Y \sim G_\omega$, then differentiation of ψ shows that

$$E_\omega(Y) = \nabla\psi(\omega) = \left[\frac{\partial}{\partial\omega_1}\psi(\omega), \cdots, \frac{\partial}{\partial\omega_p}\psi(\omega)\right]',$$

$$C_\omega(Y) = \nabla^2\psi(\omega) = \left[\frac{\partial^2}{\partial\omega_i\partial\omega_j}\psi(\omega) \colon i, j, = 1, \cdots, p\right]$$

are the mean vector and covariance matrix of Y. In particular, $\nabla^2\psi(\omega)$ is nonnegative definite, so that $\nabla^2\psi(\omega)$ is nonsingular if and only if $\nabla^2\psi(\omega)$ is positive definite (>0).

LEMMA 8.1. *If $\nabla^2\psi(\omega_0)$ is singular for some $\omega_0 \in \Omega$, then there are a proper linear subspace $L \subset R^p$ and a vector $b \in R^p$ for which $G_\omega\{b + L\} = 1$ for all $\omega \in \Omega$.*

Proof. If $\nabla^2\psi(\omega_0)$ is singular, then there is an $a \in R^p$ for which $a \neq 0$ and $\nabla^2\psi(\omega_0)a = 0$. If $Y \sim G_{\omega_0}$, then the variance of $a'Y$ is $a'\nabla^2\psi(\omega_0)a = 0$; so, $a'Y = a'\nabla\psi(\omega_0)$ w.p. 1 (G_{ω_0}). Let $b = a'\nabla\psi(\omega_0)$ and $L = \{y \in R^p \colon a'y = 0\}$. Then $G_{\omega_0}\{b + L\} = 1$; and the lemma follows from the fact that the distributions G_ω, $\omega \in \Omega$, are mutually absolutely continuous.

Thus, if $\nabla^2\psi(\omega)$ is singular for some $\omega \in \Omega$, then the family (8.1) may be parameterized with fewer parameters. In view of this, there is no loss of generality in supposing that $\nabla^2\psi(\omega)$ is positive definite for all $\omega \in \Omega$, in which case $\nabla\psi$ is a one-to-one, continuously differentiable transformation from Ω onto its range $\mathcal{Y}_0 = \nabla\psi(\Omega)$.

Composite hypotheses of the form $\omega \in \Omega_0$ are considered. There is no loss of generality in supposing that $0 \in \Omega_0$, that $\lambda = G_0$, and that $\nabla\psi(0) = 0$, for this case too may be obtained by a simple reparameterization. For simplicity, attention is restricted to linear hypotheses of the form

$$\Omega_0 = \{A\theta \colon \theta \in R^q, A\theta \in \Omega\}, \tag{8.2}$$

where $1 \leqq q < p$ and A is a $p \times q$ matrix of rank q. Below, (8.2) is to be interpreted as $\{0\}$ when $q = 0$.

For each $y \in \mathcal{Y}_0 = \nabla\psi(\Omega)$, there is a unique solution $\hat{\omega} = \hat{\omega}(y)$ to the equation $\nabla\psi(\omega) = y$, since $\nabla\psi$ is one-to-one. The function $\hat{\omega}$ is then a continuously differentiable function from $\mathcal{Y}_0$ onto Ω, by the implicit function theorem. A similar construction is possible for the subspace Ω_0. Observe first that if $Y \sim G_\omega$, $\omega \in \Omega_0$, then the distributions of $Z = A'Y$ form an exponential family with p replaced by q, ω by θ, $\psi(\omega)$ by $\psi(A\theta)$, Ω by $\Theta = \{\theta \in R^q \colon A\theta \in \Omega\}$, and $\mathcal{Y}_0$ by $\mathcal{Z}_0 = A'\mathcal{Y}_0$.

Thus, if $z \in A'\mathcal{Y}_0$, then there is a unique solution $\hat{\theta} = \hat{\theta}(z) \in \theta$ to the equation $A'\nabla\psi(A\theta) = z$. Let $\hat{\omega}_0(y) = A\hat{\theta}(A'y)$ for $y \in \mathcal{Y}_0$. Then

$$A'\nabla\psi[\hat{\omega}_0(y)] = A'y, \qquad y \in \mathcal{Y}_0,$$

and $\hat{\omega}$ satisfies this equation with A replaced by the identity. It is clear that $\hat{\omega}(y) = 0$ if and only if $y = 0$. For $\hat{\omega}_0$, one has $\hat{\omega}_0(y) = 0$ if and only if $\hat{\theta}(A'y) = 0$ if and only if $A'y = 0$. Finally, differentiation shows that

$$\omega_0^* = \hat{\omega}_0[\nabla\psi(\omega)]$$

minimizes $J[\omega_0, \omega] = (\omega - \omega_0)'\nabla\psi(\omega) - [\psi(\omega) - \psi(\omega_0)]$ among all $\omega_0 \in \Omega_0$ for each $\omega \in \Omega$.

It is easily seen that $\omega'y - \psi(\omega)$ is maximized with respect to $\omega \in \Omega$ (respectively, Ω_0) at $\omega = \hat{\omega}(y)$ (respectively, $\hat{\omega}_0(y)$) for each $y \in \mathcal{Y}_0$. Thus,

$$\phi(y) \underset{\text{def}}{=} \sup_{\Omega} [\omega'y - \psi(\omega)] = \hat{\omega}(y)'y - \psi[\hat{\omega}(y)],$$

$$\phi_0(y) \underset{\text{def}}{=} \sup_{\Omega_0} [\omega'y - \psi(\omega)] = \hat{\omega}_0(y)'y - \psi[\hat{\omega}_0(y)], \qquad y \in \mathcal{Y}_0.$$

Observe that ϕ and ϕ_0 are convex and continuously differentiable on $\mathcal{Y}_0$. In fact, $\nabla\phi(y) = \hat{\omega}(y)$ and $\nabla\phi_0(y) = \hat{\omega}_0(y)$ for $y \in \mathcal{Y}_0$. Observe also that

$$[\phi(y) - \phi_0(y)] = J[\hat{\omega}_0(y), \hat{\omega}(y)], \qquad y \in \mathcal{Y}_0.$$

Now let $Y_1, Y_2, \cdots$ be i.i.d. with common distribution G_ω for some unknown $\omega \in \Omega$, and write P_ω for probability to emphasize the dependence on ω. If $Y_1, \cdots, Y_n$ are observed, then the log-likelihood function is $l_n(\omega) = l_n(\omega; Y_1, \cdots, Y_n) = n[\omega'\bar{Y}_n - \psi(\omega)]$, $\omega \in \Omega$, where $\bar{Y}_n = (Y_1 + \cdots + Y_n)/n$. Thus, if $\bar{Y}_n \in \mathcal{Y}_0$, then the unconstrained maximum likelihood estimate is $\hat{\omega}_n =$

$\hat{\omega}(\bar{Y}_n)$, and the constrained maximum likelihood estimate is $\hat{\omega}_0(\bar{Y}_n)$. Moreover, if $\bar{Y}_n \in \mathcal{Y}_0$, then the log-likelihood ratio statistic for testing $\omega \in \Omega_0$ is

$$\Lambda_n = n[\phi(\bar{Y}_n) - \phi_0(\bar{Y}_n)] = nJ[\hat{\omega}_0(\bar{Y}_n), \hat{\omega}(\bar{Y}_n)].$$

A fixed sample size likelihood ratio test rejects $\omega \in \Omega_0$ if and only if Λ_n exceeds a given critical level. In some cases the significance levels of such tests may be estimated by using a chi-squared approximation to the null distribution of Λ_n. Here approximations based on Theorem 8.1 below are more useful.

A distribution G on R^p is said to satisfy condition S if and only if some power of its characteristic function is integrable with respect to Lebesgue measure on R^p. If G satisfies condition S and if $Y_1, Y_2, \cdots$ are i.i.d. with common distribution G, then the sum $n\bar{Y}_n = Y_1 + \cdots + Y_n$ has a bounded continuous density g_n for sufficiently large n, by Fourier inversion. Now let G_ω, $\omega \in \Omega$, be an exponential family for which G_0 satisfies condition S. Then, for sufficiently large n, the sums $n\bar{Y}_n = Y_1 + \cdots + Y_n$ have continuous densities $g_{n,\omega}$, $\omega \in \Omega$, which are related by

$$g_{n,\omega}(x) = \exp[\omega' x - n\psi(\omega)] g_{n,0}(x) \tag{8.3}$$

for all $x \in R^p$ and $\omega \in \Omega$.

THEOREM 8.1. *Suppose that* G_0 *satisfies condition* S. *Then*

$$g_{n,0}(ny) \sim \left\{\frac{1}{\sqrt{(2\pi n)^p |\nabla^2\psi(\hat{\omega}(y))|}}\right\} \exp[-n\phi(y)], \qquad y \in \mathcal{Y}_0. \tag{8.4}$$

Outline of the proof. Given $y \in \mathcal{Y}_0$, let $\hat{\omega} = \hat{\omega}(y)$. Then $g_{n,\hat{\omega}}(ny) \sim 1/\sqrt{(2\pi n)^p |\nabla^2\psi(\hat{\omega})|}$ as $n \to \infty$, by the multivariate local central limit theorem, so

$$g_{n,0}(ny) = \exp[-n\phi(y)] g_{n,\hat{\omega}}(ny)$$

is asymptotic with the right side of (8.3) as $n \to \infty$.

Remarks and references. Efron (1978) is recommended for background on the geometry of exponential families. Theorem 8.1 is taken from Borovkov and Rogozin (1965). A careful reading of their article shows that if one member of the exponential family satisfies condition S, then all do. Thus, condition S is a property of the family G_ω, $\omega \in \Omega$.

8.2. Repeated likelihood ratio tests. As in the previous section, let G_ω, $\omega \in \Omega$, denote an exponential family for which the natural parameter space Ω is open in R^p, $0 \in \Omega$, $\psi(0) = 0$, and $\nabla\psi(0) = 0$; and let $Y_1, Y_2, \cdots$ be i.i.d. with common distribution G_ω for some unknown $\omega \in \Omega$. We suppose throughout that G_0 satisfies condition S. Thus, there is an integer m_0 for which the convolutions G_0^{*n} are absolutely continuous with respect to Lebesgue measure for all $n \geqq m_0$, in which case G_ω^{*n} are absolutely continuous with respect to Lebesgue measure for all $\omega \in \Omega$ and all $n \geqq m_0$. It follows that

$$P_\omega\{\bar{Y}_n \in \mathcal{Y}_0\} = 1 \quad \text{for all } \omega \in \Omega, \quad n \geqq m_0.$$

See Barndorff-Nielson (1978, Thm. 9.2).

Now let Ω_0 be a linear hypothesis, as in (8.2). Then repeated likelihood ratio tests of Ω_0, with initial sample size $m \geqq m_0$, maximum sample size $N > m$, and (logarithmic) critical level a, reject the null hypothesis $\omega \in \Omega_0$ if and only if $\Lambda_n > a$ for some n, $m \leqq n \leqq N$. Thus, letting

$$t = t_a = \inf \{n \geqq m : \Lambda_n > a\}, \tag{8.5}$$

the tests take $T_a = \min(t, N)$ observations and reject the null hypothesis if and only if $t \leqq N$. The power function is, therefore,

$$\beta(\omega) = P_\omega\{t \leqq N\}, \qquad \omega \in \Omega.$$

Approximations to the power function and expected sample size are developed below, under the assumption that condition S is satisfied. For the sake of brevity, neither complete generality nor complete rigor is attempted.

Several previous results admit straightforward extensions to the multi-parameter context. For example, recall that $\Lambda_n = n[\phi(\bar{Y}_n) - \phi_0(\bar{Y}_n)] = nJ[\hat{\omega}_0(\bar{Y}_n), \hat{\omega}(\bar{Y}_n)]$ for $n \geqq m_0$ and observe that for fixed $\omega \in \Omega - \Omega_0$,

$$\Lambda_n = S_n + \xi_n,$$

where

$$S_n = S_n(\omega) = nJ(\omega_0^*, \omega) + n(\omega - \omega_0^*)'[\bar{Y}_n - \nabla\psi(\omega)], \qquad n \geqq 1,$$

with

$$\omega_0^* = \hat{\omega}_0[\nabla\psi(\omega)],$$

and

$$\xi_n = \xi_n(\omega) = n[\bar{Y}_n - \nabla\psi(\omega)]'M_n[\bar{Y}_n - \nabla\psi(\omega)],$$

with

$$M_n = \int_0^1 \nabla^2(\phi - \phi_0)[s\bar{Y}_n + (1-s)\nabla\psi(\omega)](1-s)\,ds, \qquad n \geqq m_0.$$

As in Lemmas 6.3 and 6.4, it may shown that ξ_n, $n \geqq m_0$, are slowly changing with respect to P_ω for all $\omega \in \Omega - \Omega_0$ and that $S_1(\omega)$ has a nonarithmetic distribution for almost every $\omega \in \Omega - \Omega_0$. Observe that the random walks $S_n = S_n(\omega)$ have positive drifts. Let H_ω denote the asymptotic distribution of residual waiting time for $S_n = S_n(\omega)$, $n \geqq 1$. Then $\Lambda_t - a$ has asymptotic distribution H_ω, under P_ω, for almost every $\omega \in \Omega$. Let $\mathscr{H}(\omega, \cdot)$ denote the Laplace transform of H_ω, $\omega \in \Omega$.

Similarly, letting

$$u_a(n, y) = P\{t \geqq n \mid \bar{Y}_n = y\}, \qquad y \in \mathscr{Y}_0, \quad n \geqq 1, \quad a > 0,$$

one finds: if $n = n_a \to \infty$ and $y_a \to y \in \mathscr{Y}_0$ as $a \to \infty$, with $n[\phi(y_a) - \phi_0(y_a)] \to r > 0$, then

$$u_a(n, y_a) \to v[\hat{\omega}(y), r],$$

where

$$v[\omega, r] = P_\omega\{S_k(\omega) \geqq r, \text{ for all } k \geqq 1\}, \qquad r>0, \quad \omega \in \Omega.$$

THEOREM 8.2. *Let G_ω, $\omega \in \Omega$, be an exponential family for which the natural parameter space Ω is open in R^p, $0 \in \Omega$, $\psi(0)=0$, and $\nabla\psi(0)=0$, and suppose that G_0 satisfies condition S. Let Ω_0 be a linear hypothesis, as in (8.2), let $l=p-q$, and define $t=t_a$ by (8.5). If $m=m_a \to \infty$ and $N=N_a \to \infty$ as $a \to \infty$ with $a/N \to \delta_0$ and $a/m \to \delta_1$, where $0<\delta_0<\delta_1<\infty$, then*

$$P_0\{t \leqq N\} \sim K a^{l/2} e^{-a} \quad \text{as } a \to \infty,$$

where

$$K = C_A \int_{\substack{\delta_0 < J[0,\hat\omega] < \delta_1 \\ \hat\omega_0 = 0}} \frac{\mathcal{H}(\hat\omega, 1)}{\sqrt{J(0,\hat\omega)^l}\,|\nabla^2\psi(\hat\omega)|}\, dS^l(y),$$

$$C_A = |A'\nabla^2\psi(0)A|^{1/2}/\sqrt{2\pi}^{\,p-q}|A'A|^{1/2},$$

S^d denotes d-dimensional surface area in R^p, $|\cdot|$ denotes determinant, and $\mathcal{H}(\omega, \cdot)$ denotes the Laplace transform of the asymptotic distribution of residual waiting time for the random walk $S_k(\omega)$, $k \geqq 1$.

Outline of the proof. First consider $n=n_a$ for which $\varepsilon = a/n$ converges to an interior point of (δ_0, δ_1) as $a \to \infty$. Then Theorem 8.1 suggests

$$P_0\{t=n\} = \int_{\phi-\phi_0>\varepsilon} u_a(n, y) n^p g_{n,0}(ny)\, dy$$

$$\sim \sqrt{\left(\frac{n}{2\pi}\right)^p} \int_{\phi-\phi_0>\varepsilon} u_a(n,y)|\nabla^2\psi(\hat\omega)|^{-1/2} \exp\{-n\phi(y)\}\, dy \tag{8.6}$$

$$= \sqrt{\left(\frac{n}{2\pi}\right)^p} \int_\varepsilon^\infty I_n(\varepsilon, s)\, e^{-ns}\, ds,$$

where

$$I_n(\varepsilon, s) = \int_{\phi=s, \phi-\phi_0>\varepsilon} \left\{\frac{u_a(n,y)}{\|\hat\omega\| \cdot |\nabla^2\psi(\hat\omega)|^{1/2}}\right\} dS^{p-1}(y). \tag{8.7}$$

Clearly, the main contribution to the final integral in (8.6) comes from s for which $s=\varepsilon + O(1/n)$. Then one has $n(\phi-\phi_0)(y) = n\varepsilon + O(1)$ and $\phi_0(y) = O(1/n)$ in (8.7), suggesting

$$u_a(n, y) \approx v[\hat\omega(y), r],$$

where $r = n(\phi - \phi_0)(y)$.

When these substitutions are combined with some simple analysis, the change of variables $s' = ns$ and an interchange of orders of integration, one finds that

$$P_0\{t=n\} \sim c(\varepsilon) a^{l/2-1} e^{-a} \quad \text{as } a \to \infty, \tag{8.8}$$

where

$$c(\varepsilon)=\int_{\phi_0=0,\phi=\varepsilon}\mathscr{H}(\hat{\omega},1)\,dS_0(y)$$

and

$$dS_0(y)=C_A\sqrt{\left(\frac{1}{2\pi}\right)^l}\,\varepsilon^{-l/2+2}\{\|\hat{\omega}\|\cdot|\nabla^2\psi(\hat{\omega})|\}^{-1/2}\,dS^{l-1}(y).$$

Finally, if (8.8) is substituted into the relation $P_0\{t\leqq N\}=\sum_{n=m}^{N}P_0\{t=n\}$ and if a summation is compared to a Riemann integral, then one finds that

$$P_0\{t\leqq n\}\sim Ka^{l/2}e^{-a}\quad\text{as } a\to\infty,$$

(8.9)
$$K=\int_{\delta_0}^{\delta_1}c(\varepsilon)\varepsilon^{-2}\,d\varepsilon,$$

and this is equivalent to the assertion of the theorem.

By way of contrast, the extension of approximations to the expected sample size from the context of one parameter to that of several presents only notational difficulties. Recall that the sample size is $T=T_a=\min(t_a,N)$; and let

$$\mu_\omega=J[\omega_0^*,\omega],\quad \sigma_\omega^2=\operatorname{tr}\{\nabla^2\psi(\omega)\cdot\nabla^2(\phi-\phi_0)(\nabla\psi(\omega))\},\quad \rho_\omega=\int_0^\infty rH_\omega\{dr\},$$

where H_ω denotes the asymptotic distribution of residual waiting time for the random walk $S_k(\omega)$, $k\geqq 1$.

THEOREM 8.3. *Under the assumptions of Theorem* 8.2, *as* $a\to\infty$,

$$E_\omega(T_a)=\begin{cases}N+o(1), & \mu_\omega<\delta_0,\\ (1/\mu_\omega)[a+\rho_\omega-\frac{1}{2}\sigma_\omega^2]+o(1), & \delta_0<\mu_\omega<\delta_1,\\ m+o(1), & \mu_\omega>\delta_1.\end{cases}$$

Remarks and references. Theorem 8.2 is adapted from Woodroofe (1978). Additional details may be found there. The conditions imposed in Theorem 8.2 are clearly stronger than are necessary, for (8.8) is stronger than the assertion of the theorem. Lalley (1980) has recently obtained a substantial generalization of Theorem 8.2, without the smoothness condition S.

8.3. Examples. The computational problems are harder with several parameters, but not insurmountable. Computational aspects are illustrated with two examples.

Example 8.1. *Gamma distributions.* Let Z_1 and Z_2 be independent with marginal densities

$$k_i(z;\theta_i)=\frac{1}{\theta_i^s\Gamma(s)}z^{s-1}\exp\left(-\frac{z}{\theta_i}\right),\qquad z>0,$$

where $-\infty<\theta_1,\theta_2<\infty$ are unknown and $s>0$ is known. Then repeated likelihood ratio tests may be used to test the hypothesis $\theta_1=\theta_2$ using independent replications of $Z = (Z_1, Z_2)'$. Observe that this formulation includes the problems of comparing two normal variances or two exponential failure rates as special cases.

Considerations of symmetry show that the probability of a type I error is independent of θ, when $\theta_1 = \theta_2 = \theta$. We use Theorem 8.2 to compute it when $\theta_1 = \theta_2 = 1$. The transformations

$$Y_i = \frac{1}{s} Z_i - 1 \quad \text{and} \quad \omega_i = s\left[1 - \frac{1}{\theta_i}\right], \qquad i = 1, 2$$

reduce the problem to the form considered in § 8.2 with $\Omega = (-\infty, s)^2$ and $\mathscr{Y}_0 = (-1, \infty)^2$. Moreover, $\omega_1 = \omega_2 = 0$ when $\theta_1 = \theta_2 = 1$. The unrestricted maximum likelihood estimates are $\hat{\theta}_i = z_i/s$, or $\hat{\omega}_i = sy_i/(1+y_i)$, $i = 1, 2$, and the restricted maximum likelihood estimates are $\hat{\theta}_{01} = \hat{\theta}_{02} = (z_1 + z_2)/2s$, or $\hat{\omega}_{01} = \hat{\omega}_{02} = (y_1 + y_2)s/(2 + y_1 + y_2)$. Thus,

$$(\phi - \phi_0)(y_1, y_2) = 2s \log\left[\frac{z_1 + z_2}{2s}\right] - s \log\left[\frac{z_1 z_2}{s^2}\right]$$

$$= 2s \log\left[1 + \frac{y_1 + y_2}{2}\right] - s \log\left[(1+y_1)(1+y_2)\right].$$

For $\varepsilon > 0$, $M_\varepsilon = \{y : \hat{\omega}_0(y) = 0 \text{ and } \phi(y) = \varepsilon\}$ is the two-point set

$$y_1 = -y_2 = \pm\sqrt{1 - e^{-\varepsilon/s}} = \pm b, \quad \text{say},$$

so

$$\text{(8.10)} \qquad \hat{\omega}_1 = \pm \frac{bs}{1 \pm b} \quad \text{and} \quad \hat{\omega}_2 = \mp \frac{bs}{1 \mp b}$$

for $y \in M_\varepsilon$. It is shown below that $\mathscr{H}(\hat{\omega}, 1)$ has the same value for both such ω, and when this remark is combined with some simple algebra one finds

$$c(\varepsilon) = \varepsilon^{3/2}[\pi s b^2(1 + b^2)]^{-1/2}\mathscr{H}(\hat{\omega}, 1), \qquad \varepsilon > 0,$$

in (8.8) and (8.9). Thus, the problem is to compute $\mathscr{H}(\hat{\omega}, 1)$.

When ω is as in (8.10), one finds that $\theta_1 - 1 = \pm b$ and $\theta_2 - 1 = \mp b$, and

$$X = \varepsilon \pm b\left[\frac{Z_1}{\theta_1} - \frac{Z_2}{\theta_2}\right] = \varepsilon \pm b(G_1 - G_2),$$

where G_1 and G_2 are independent gamma random variables with shape parameters s and unit scale parameters. Thus the characteristic function of X is

$$E[e^{itX}] = (1 + b^2 t^2)^{-s} e^{it\varepsilon}, \qquad -\infty < t < \infty,$$

and $\mathscr{H}$ may be computed from the inversion formula, Theorem 2.6, by straightforward numerical integration.

In the special case that $s=1$, X has an exponential left tail; and one may use this remark to compute $\mathcal{H}(\omega, 1)$ explicitly as

$$\mathcal{H}(\omega, 1)=[1+\sqrt{1-e^{-\varepsilon}}]^{-1}\left[\frac{2}{\varepsilon}(1-e^{-\varepsilon})-e^{-\varepsilon}\right], \qquad \varepsilon>0.$$

Example 8.2. *Repeated t-tests.* Suppose that Z has normal distribution with unknown mean θ, $-\infty<\theta<\infty$, and unknown variance $\sigma^2>0$, and that one wishes to test $\theta=0$ on the basis of independent replications of Z. Then the type I error probability of the repeated likelihood ratio tests is independent of $\sigma^2>0$ when $\theta=0$, and may be computed from Theorem 8.2 when $\theta=0$ and $\sigma^2=1$.

The density of Z with respect to the standard normal distribution is

$$\frac{1}{\sigma}\exp\left[-\frac{(z-\theta)^2}{2\sigma^2}+\frac{z^2}{2}\right]=\exp\left[\omega_1 z^2+\omega_2 z-\psi(\omega_1, \omega_2)\right],$$

where

$$\omega_1=\frac{1-\sigma^{-2}}{2}, \quad \omega_2=\sigma^{-2}\theta, \quad \psi=\frac{\sigma^{-2}\theta^2+\log\sigma^2}{2}.$$

Thus the distributions of $Y=(Z^2, Z)'$ are of the form (8.1) with $\Omega=\{\omega\colon \omega_1<\frac{1}{2}, -\infty<\omega_2<\infty\}$ and $\mathcal{Y}_0=\{(y_1, y_2)\colon y_2^2<y_1\}$. The unrestricted maximum likelihood estimates are $\hat{\theta}=y_2$ and $\hat{\sigma}^2=y_1-y_2^2$, the restricted maximum likelihood estimates are $\hat{\theta}_0=0$ and $\hat{\sigma}_2^2=y_1$ and

$$(\phi-\phi_0)(y_1, y_2)=\frac{\log y_1-\log(y_1-y_2^2)}{2}.$$

For $\varepsilon>0$, $M=\{y\colon \hat{\omega}_0(y)=0, \phi(y)=\varepsilon\}$ is the two-point set $y_1=1$ and $y_2=\pm\sqrt{1-e^{-2\varepsilon}}$, and

$$\text{(8.11)} \qquad \hat{\omega}_1=\frac{1-e^{-2\varepsilon}}{2} \quad \text{and} \quad \hat{\omega}_2=\pm e^{2\varepsilon}\sqrt{1-e^{-2\varepsilon}}$$

for $y\in M_\varepsilon$. As above, $\mathcal{H}(\hat{\omega}, 1)$ has the same value for both such values of $\hat{\omega}$ and, when this remark is combined with some straightforward algebra, one finds

$$c(\varepsilon)=\varepsilon^{3/2}\left(\frac{2}{\pi[\sinh^2(\varepsilon)+(e^{2\varepsilon}-1)]}\right)e^{2\varepsilon}\mathcal{H}(\hat{\omega}, 1), \qquad \varepsilon>0,$$

in (8.8) and (8.9). To find $\mathcal{H}$ we must find the distribution of $X=\varepsilon+\omega_1(Z^2-1)+\omega_2(Z-\theta)$, when ω is as in (8.11). Simple algebra shows that

$$X=\frac{(1+2\varepsilon)}{2}-\frac{\theta^2}{2}\left(W-\frac{\sigma}{\theta}\right)^2,$$

where $W=(Z-\theta)/\sigma$ has the standard normal distribution, so X has the distribution of $(1+2\varepsilon)/2-\theta^2\chi_1^2(\theta^{-2}\sigma^2)/2$. It follows easily that

$$E[e^{itX}]=[1+i\theta^2 t]^{-1/2}\exp\left[-(1+i\theta^2 t)\sigma^2 t+(1+2\varepsilon)it\right], \qquad -\infty<t<\infty,$$

from which $\mathcal{H}$ may be computed by inversion.

Remarks and references. Examples 8.1 and 8.2 are taken from Woodroofe (1979). The problem of sequentially testing that a correlation coefficient is zero is also considered there and brief tables for the constant K are given for these three examples.

Siegmund (1980) has developed repeated significance tests for one-way analysis of variance models, including power calculations and confidence regions as well as significance levels.

CHAPTER 9

Estimation Following Sequential Testing

9.1. The bias. As demonstrated in Chapter 7, repeated likelihood ratio tests offer the possibility of substantial savings in the number of observations, when the parameter is far from the null hypothesis. Suppose now that one uses repeated likelihood ratio tests to test a given hypothesis, say $\omega = 0$. Then, in addition to testing the null hypothesis, one might want to estimate ω or the mean $\psi'(\omega)$, especially if the null hypothesis is rejected. And here the use of sequential methods introduces some additional complications, for they change the sampling distribution of the sufficient statistics from their familiar fixed sample size values and introduce some additional biases into the maximum likelihood estimates.

In this chapter the problems of point and interval estimation following sequential testing are considered. For simplicity, attention is restricted to the normal case. Thus, let $Y_1, Y_2, \cdots$ denote i.i.d. normally distributed random variables with unknown mean θ, $-\infty < \theta < \infty$, and unit variance, and consider tests of the null hypothesis $\theta = 0$. It may be convenient to think of Y_i as the difference between the response to an experimental treatment and a placebo by a typical subject or pair of subjects, in which case θ is the average difference which would be observed if the treatments were given to infinitely many (pairs of) subjects. Let

$$t = t_a = \inf\left\{n \geqq 1: \frac{n\bar{Y}_n^2}{2} > a\right\}, \qquad a > 0.$$

Then repeated likelihood ratio tests with initial sample size $m = 1$, maximum sample size N, and critical levels $a = b$ take $T = \min(t, N)$ observations and reject $\theta = 0$ if and only if $t \leqq N$. The power function of this test may be approximated from the results of §§ 7.2 and 7.4.

For purposes of estimation, one observes $T = \min(N, t)$ and $Y_1, \cdots, Y_T$ and wants to estimate θ. The relevant distributions are then the restrictions Q_θ of P_θ to $\mathscr{A}_T$, the sigma-algebra of events prior to T, and

$$\frac{dQ_\theta}{dQ_0} = \exp\left\{\theta s_T - \frac{1}{2}\theta^2 T\right\}, \qquad -\infty < \theta < \infty,$$

where

$$s_T = T\bar{Y}_T = Y_1 + \cdots + Y_T.$$

In the terminology of Efron (1978), $T, Y_1, \cdots, Y_T$ form a sample of size one from a curved exponential family. The likelihood function is unaffected by optional stopping, so the maximum likelihood estimator is $\hat{\theta} = \bar{Y}_T$. But the sampling distribution of $\bar{Y}_T$ is affected by the optional stopping. The first theorem illustrates this fact by computing the bias of $\hat{\theta} = \bar{Y}_T$.

THEOREM 9.1. *Suppose that* $N = [a\delta_0^{-1}]$, *where* $0 < \delta_0 < \infty$, *and let* $\theta_0 = \sqrt{2\delta_0}$. *Then*

$$E_\theta(\bar{Y}_T) = \begin{cases} \theta + o\left(\dfrac{1}{a}\right), & |\theta| < \theta_0 \\ \theta\left[1 + \dfrac{1}{a}\right] + o\left(\dfrac{1}{a}\right), & |\theta| > \theta_0 \end{cases} \quad \textit{as } a \to \infty.$$

Proof. It suffices to consider $\theta \geqq 0$. If $0 \leqq \theta < \theta_0$, then there are C and γ for which

$$P_\theta\{t < N\} \leqq Ce^{-\gamma a}, \qquad a > 0,$$

so

$$\begin{aligned} |E_\theta(\bar{Y}_T - \theta)| = |E_\theta(\bar{Y}_T - \bar{Y}_N)| &\leqq E_\theta[|\bar{Y}_T - \bar{Y}_N|] \\ &\leqq \sqrt{E_\theta(\bar{Y}_T - \bar{Y}_N)^2 P_\theta\{t < N\}} \\ &\leqq 2\sqrt{E_\theta[\sup_n \bar{Y}_n^2] \times Ce^{-\gamma a}} = o\left(\frac{1}{a}\right), \qquad a \to \infty. \end{aligned}$$

For $\theta > \theta_0$, $T/a \to 2/\theta^2 \leftarrow t/a$ w.p.1 (P_θ). Now

$$(\bar{Y}_t - \theta) = t^{-1}(s_t - t\theta) = \frac{\theta^2}{2a}(s_t - t\theta) + \left(1 - \frac{t\theta^2}{2a}\right)[\bar{Y}_t - \theta].$$

The expectation of the first term on the right is zero by Wald's lemma. To understand the second, write

$$\frac{t\bar{Y}_t^2}{2} = a + R_a$$

and

$$\left(1 - \frac{\theta^2 t}{2a}\right) = \bar{Y}_t^{-2}[\bar{Y}_t^2 - \theta^2] + O_p\left(\frac{1}{a}\right) \quad \text{as } a \to \infty.$$

Then

$$\begin{aligned} W_a \underset{\text{def}}{=} a\left(1 - \frac{\theta^2 t}{2a}\right)(\bar{Y}_t - \theta) \\ = a\bar{Y}_t^{-2}(\bar{Y}_t + \theta)(\bar{Y}_t - \theta)^2 + o_p(1) \Rightarrow \theta\chi_1^2 \quad \text{as } a \to \infty. \end{aligned}$$

So, it would suffice to verify that W_a, $a > 0$, are uniformly integrable. The details of this verification have been omitted.

Remarks and references. Theorem 9.1 was conjectured by Cox (1952) and proved by Siegmund (1978).

9.2. Confidence intervals. The asymptotic distribution of $\sqrt{T}(\bar{Y}_T - \theta)$ is normal with mean 0 and unit variance, by Anscombe's theorem, so asymptotic confidence intervals may be set quite simply. For example, the interval $\bar{Y}_T \pm 1.96T^{-1/2}$ is an asymptotic 95% confidence interval for θ. On the other hand, Theorem 9.1 indicates that the approximation by Anscombe's theorem may be inadequate in some cases, since the bias $E_\theta[\bar{Y}_T-\theta]\approx\theta/a$ may be appreciable for, for example, $a=5$ and $\theta=2$. In this section an alternative method for setting confidence limits is developed. The alternative uses the well known relation between confidence sets and test of hypotheses. See, for example, Lehmann (1959, §§ 3.5 and 5.4). It is reviewed in the next paragraph.

Denote the desired confidence coefficient by $1-2\alpha$; and suppose that $2\alpha > \alpha^* = P_0\{t_a \leqq N\}$, the significance level of the test. For each ξ, $-\infty<\xi<\infty$, and sufficiently large a, a level α test of the hypothesis $\theta\leqq\xi$ versus $\theta>\xi$ is constructed below. Let $A(\xi)$ denote the acceptance region of this test, in the space of $(T, Y_1, \cdots, Y_T)$. Thus,

$$P_\theta\{(T, Y_1, \cdots, Y_T)\in A(\xi)\}\geqq 1-\alpha \quad \text{for all } \theta\leqq\xi.$$

Then

$$S(T, Y_1, \cdots, Y_T)=\{\theta\colon (T, Y_1, \cdots, Y_T)\in A(\theta)\}$$

is a level $1-\alpha$ confidence set for θ, since $P_\theta\{\theta\in S(T, Y_1, \cdots, Y_T\}= P_\theta\{(T, Y_1, \cdots, Y_T)\in A(\theta)\}\geqq 1-\alpha$ for all θ. The acceptance regions $A(\xi)$ below are increasing in that $A(\xi_1)\subset A(\xi_2)$ when $\xi_1<\xi_2$. It follows that $S(T, Y_1, \cdots, Y_T)$ are of the form

$$S(T, Y_1, \cdots, Y_T)=(\theta_*, \infty),$$

where $\theta_* = \theta_*(T, Y_1, \cdots, Y_T)$.

Let

$$\theta^* = \theta_*(T, -Y_1, \cdots, -Y_T).$$

Then (θ_*, θ^*) is a confidence interval with confidence coefficient $1-2\alpha$, by symmetry. Explicit expressions for θ_* and θ^* are given below.

The family of tests are defined by giving their critical regions, the complements of the acceptance regions. It is assumed throughout that

$$N=N(a)=a\delta_0^{-1}, \qquad a>0, \tag{9.1}$$

where $0<\delta_0<\infty$ remains fixed as $a\to\infty$. First consider testing $\theta\leqq\xi$, where $\xi<0$. For $1\leqq n\leqq N$, let C_n^- be the event $C_n^-=\{T<n, \bar{Y}_T>0\}\cup\{T\geqq n\}$. Then the test which rejects $\theta\leqq\xi$ if and only if C_n^- occurs has power function

$$\begin{aligned}\beta_n^-(\theta)&=P_\theta\{t<n, \bar{Y}_t>0\}+P_\theta\{t\geqq n\}\\&=P_\theta\{\bar{Y}_t>0\}+P_\theta\{t\geqq n, \bar{Y}_t\leqq 0\}.\end{aligned}$$

Observe that β_n^- is continuous and increasing in θ for fixed n, by Theorem 3.5, and β_n^- is decreasing in n for fixed θ. Moreover, $\beta_n^-(\theta)\to 0$ or 1 as $\theta\to-\infty$ or

$\theta\to\infty$ for $n\geqq 2$. Let $1-2\alpha$ denote the desired confidence coefficient and let

$$\xi^-(n)=\inf\{\theta: \beta_n^-(\theta)\geqq\alpha\},\qquad 1\leqq n\leqq N.$$

Then $\xi^-(1)=-\infty$, since $\beta_1^-(\theta)=1$ for all θ; $\xi^-(2)<\cdots<\xi^-(N)$; and $\xi^-(2)\to-\infty$, as $a\to\infty$, since $\beta_2^-(\theta)\to 1$ as $a\to\infty$ for each fixed θ. Moreover, $\beta_N^-(0)\to 1$ as $a\to\infty$, so that $\xi^-(N)<0$ for all sufficiently large a. Next, let $C(\xi)=C_n^-$ for $\xi^-(n-1)<\xi\leqq\xi^-(n)$ and $n=2,\cdots,N$. Then $P_\theta[C(\xi)]\leqq P_\xi[C(\xi)]\leqq\alpha$ for all $\theta\leqq\xi$ and $\xi\leqq\xi^-(N)$, so $C(\xi)$ is the critical region of a level α test of $\theta\leqq\xi$ for $\xi\leqq\xi^-(N)$. If $\xi^-(n-1)<\xi\leqq\xi^-(n)$, then the acceptance region $A(\xi)=C(\xi)'=C_n^{-\prime}$ may be written $A(\xi)=\{T<n,\ \bar{Y}_T\leqq 0\}$; and, since $T<n$ if and only if $\xi^-(T)<\xi^-(n)$, this may be written

$$A(\xi)=\{\xi^-(T)<\xi,\ \bar{Y}_T\leqq 0\},\qquad \xi\leqq\xi^-(N). \tag{9.2}$$

Next consider $\xi>0$. Let C_n^+ be the event $\{t\leqq n,\ \bar{Y}_t>0\}$ for $1\leqq n\leqq N$. Then the test which rejects $\theta\leqq\xi$ if and only if C_n^+ occurs has power

$$\beta_n^+(\theta)=P_\theta\{t\leqq n,\ \bar{Y}_t>0\},\qquad -\infty<\theta<\infty.$$

It is easily seen that $\beta_n^+(\theta)$ is continuous and increasing in θ for fixed n, and increasing in n for fixed θ. Moreover, $\beta_n(\theta)\to 0$ or 1 as $\theta\to-\infty$ or $\theta\to\infty$. Next, let

$$\xi^+(n)=\inf\{\theta: \beta_n^+(\theta)\geqq\alpha\},\qquad 1\leqq n\leqq N.$$

Then $\xi^+(N)<\xi^+(N-1)<\cdots<\xi^+(1)$; $\xi^+(1)\to\infty$ as $a\to\infty$, since $\beta_1^+(\theta)\to 0$ for each fixed θ; and $\xi^+(N)>0$ for all sufficiently large a, since $\beta_N^+(0)=\alpha^*/2<\alpha$, by assumption. Next, let $C(\xi)=C_n^+$ for $\xi^+(n+1)<\xi\leqq\xi^+(n)$ and $1\leqq n<N$. Then $C(\xi)$ is the critical region of a level α test of $\theta\leqq\xi$ for all ξ, $\xi^+(N)<\xi\leqq\xi^+(1)$. If $\xi^+(n+1)<\xi\leqq\xi^+(n)$, then the acceptance $A(\xi)=C(\xi)'$ may be written $A(\xi)=\{T\leqq n,\ \bar{Y}_T\leqq 0\}\cup\{T>n\}$ and, since $T>n$ if and only if $\xi^+(T)<\xi^+(n)$,

$$A(\xi)=\{T\leqq n,\ \bar{Y}_T\leqq 0\}\cup\{\xi^+(T)<\xi\} \tag{9.3}$$

for

$$\xi^+(n+1)<\xi\leqq\xi^+(n)\quad\text{and}\quad n=1,\cdots,N-1.$$

The interval $(\xi^-(N),\xi^+(N)]$ requires special treatment. For $-\infty<b<\infty$, let D_b be the event $\{T<N,\ \bar{Y}_T>0\}\cup\{T=N,\ \bar{Y}_N>b\}$. Then the test which rejects $\theta\leqq\xi$ if and only if D_b occurs has power function

$$\beta_b(\theta)=P_\theta\{T<N,\ \bar{Y}_T>0\}+P_\theta\{T=N,\ \bar{Y}_N>b\}.$$

Observe that β_b is continuous and increasing in θ for fixed b, and decreasing in b for fixed θ. Moreover, $\beta_b(\theta)\to\beta_N^-(\theta)$ as $b\to-\infty$, and $\beta_b(\theta)=\beta_N^+(\theta)$ for all sufficiently large b. Let

$$\xi_0(b)=\inf\{\theta: \beta_b(\theta)\geqq\alpha\},\qquad -\infty<b<\infty.$$

Then $\xi_0(b)$ is an increasing function of b, $\xi_0(b)\to\xi^-(N)$ as $b\to-\infty$, and $\xi_0(b)=\xi^+(N)$ for all sufficiently large b. Let $b(\xi)$ be the inverse function to $\xi_0(b)$, defined on the interval $\xi^-(N)<\xi \leqq \xi^+(N)$, and let $D(\xi)=D_{b(\xi)}$ for $\xi^-(N)<\xi\leqq\xi^+(N)$. Then $D(\xi)$ is the critical region of a level α test of $\theta\leqq\xi$ for $\xi^-(N)<\xi\leqq\xi^+(N)$. The acceptance region $A(\xi)=D(\xi)'$ may be written

$$A(\xi)=\{T<N,\ \bar{Y}_T\leqq 0\}\cup\{T=N,\ \xi_0(\bar{Y}_N)<\xi\}. \tag{9.4}$$

The confidence set $S=S(T, Y_1, \cdots, Y_T)=\{\xi: (T, Y_1, \cdots, Y_T)\in A(\xi)\}$ is now easily determined. If $T<N$ and $\bar{Y}_T>0$, then $A(\xi)$ occurs if and only if $\xi^+(T)<\xi$; if $T=N$, then $A(\xi)$ occurs if and only if $\xi_0(\bar{Y}_T)<\xi$; and if $T<N$, and $\bar{Y}_T\leqq 0$, then $A(\xi)$ occurs if and only if $\xi^-(T)<\xi$. Thus, $S=S(T, Y_1, \cdots, Y_T)$ is a interval with upper endpoint ∞ and lowe endpoint

$$\theta_* = \begin{cases} \xi^+(T), & T<N, \quad \bar{Y}_T>0, \\ \xi_0(\bar{Y}_N), & T=N, \\ \xi^-(T), & T<N, \quad \bar{Y}_T\leqq 0. \end{cases}$$

That is, $\theta_*=\theta_*(T, Y_1, \cdots, Y_T)$ is a lower confidence bound with confidence coefficient at $1-\alpha$. An upper confidence bound may now be easily constructed by letting

$$\theta^* = -\theta_*(T, -Y_1, \cdots, -Y_T),$$

and it follows that (θ_*, θ^*) is a confidence interval with confidence coefficient at least $1-2\alpha$.

The power functions β_n^-, β_n^+, and β_b may be approximated from the results of §§ 7.2 and 7.4.

Example 9.1. Suppose that repeated likelihood ratio tests of $\theta=0$ are performed with $a=b=5.95$, $m=1$ and $N=148$, yielding $\alpha^*\approx .01$. How should 90% confidence intervals be set when one observes $t=n<N$? For example, if one observes $t=10$ and $\bar{Y}_t>0$, one may approximate β_{10}^+ from Theorem 7.3 and solve the equations $\beta_{10}^+(\theta)=.05$ numerically as $\theta_*\approx .49$. If one observes $t=10$ and $\bar{Y}_t<0$, then one may solve $\beta_{10}^-(\theta)=.05$ for $\theta_*\approx -1.58$. Thus, if one observes $t=10$ and $\bar{Y}_t>0$, then (.49, 1.58) is a 90% confidence interval.

Remarks and references. The results of this section are taken from Siegmund (1978). See Siegmund (1980) for further developments and Armitage (1958) for a closely related proposal.

CHAPTER 10

Sequential Estimation

10.1. Point estimation. Let $Y_1, Y_2, \cdots$ be i.i.d. normally distributed random variables with unknown mean μ, $-\infty<\mu<\infty$, and unknown variance $\sigma^2>0$, and let

$$\hat{\mu}_n=\bar{Y}_n, \qquad \hat{\sigma}_n^2=\frac{1}{n-1}\sum_{i=1}^{n}(Y_i-\bar{Y}_n)^2, \qquad n\geqq 2$$

denote the unbiased estimators of μ and σ^2. Consider the problem of estimating the mean μ with loss $|\hat{\mu}-\mu|^{2p}$ for estimation error and cost $c>0$ for each observation, where $p>0$. If a sample of fixed size n were taken, and if μ were estimated by $\hat{\mu}_n=\bar{Y}_n$, then the risk would be

$$R_{\sigma,c}(n)=E_{\mu,\sigma}[|\bar{Y}_n-\mu|^{2p}+cn]=K_p\sigma^{2p}n^{-p}+cn,$$

where

$$K_p=\frac{2^p\Gamma(\frac{1}{2}+p)}{\sqrt{\pi}};$$

and, if σ^2 were known, then the risk would be minimized by letting n be an integer adjacent to

$$a=\left(\frac{pK_p\sigma^{2p}}{c}\right)^{1/(p+1)}.$$

Note that

$$R_{\sigma,c}(a)=\left(1+\frac{1}{p}\right)ac, \qquad c,\sigma>0.$$

For unknown σ^2, Robbins (1959) suggested the sequential procedure which estimates σ^2 at each stage n and stops as soon as $n>\hat{a}_n=(c^{-1}pK_p\hat{\sigma}_n^{2p})^{1/(p+1)}$. This procedure underestimates a slightly at the termination, so consider the simple modification which continues sampling until $n>\hat{a}_n l_n$, where $l_n>1$, $n\geqq 1$, and $l_n\to 1$ as $n\to\infty$. Letting $m\geqq 2$ be the initial sample size, the stopping time of this procedure may be written

$$T=\inf\{n\geqq m: n>bl_n\hat{\sigma}_n^{2\alpha}\}, \tag{10.1}$$

where

$$\alpha=\frac{p}{p+1} \quad\text{and}\quad b=(c^{-1}pK_p)^{1/(p+1)}.$$

Let $R^*_{\sigma,c}$ be the risk incurred by the sequential procedure and let $\mathcal{R}(\sigma, c)$ be the regret, the additional risk incurred by using T instead of a. Thus,

$$R^*_{\sigma,c} = E_{\mu,\sigma}[|\bar{Y}_T - \mu|^{2p} + cT], \qquad \mathcal{R}(\sigma,c) = R^*_{\sigma,c} - R_{\sigma,c}(a), \qquad \sigma, c > 0.$$

The main result of the section shows that $\mathcal{R}(\sigma, c)$ is of the order of magnitude of c, the cost of a single observation, as $c \to 0$.

The analysis of T is simplified by the transformation

$$W_k = \left[\sum_{j=1}^{k} (Y_j - Y_{k+1})\right]^2 \Big/ k(k+1)\sigma^2, \qquad k \geqq 1.$$

LEMMA 10.1. *$W_1, W_2, \cdots$ are i.i.d. χ_1^2 random variables, $W_1, \cdots, W_{n-1}$ are independent of $\bar{Y}_n$ for all $n \geqq 2$ and*

$$\sum_{i=1}^{n} (Y_i - \bar{Y}_n)^2 = \sigma^2 \sum_{j=1}^{n-1} W_j \quad \textit{for all } n \geqq 2.$$

Proof. Letting ε_k denote the sign of $(Y_1 + \cdots + Y_k - kY_{k+1})$, $k \geqq 1$, one finds that $[\varepsilon_1\sqrt{W_1}, \cdots, \varepsilon_{n-1}\sqrt{W_{n-1}}, \sqrt{n}/\sigma(\bar{Y}_n - \mu)]$ is an orthogonal transformation of $(Y_1 - \mu, \cdots, Y_n - \mu)/\sigma$ for each $n \geqq 2$. The lemma follows immediately.

The lemma has several important consequences. First, $\hat{\sigma}_n^2 = \sigma^2 \bar{W}_{n-1}$ for $n \geqq 2$. So, the event $\{T = n\}$ is independent of $\bar{Y}_n$, since it is determined by $\hat{\sigma}_2^2, \cdots, \hat{\sigma}_n^2$, for $n \geqq 2$. It follows that

$$E_{\mu,\sigma}[|\bar{Y}_n - \mu|^{2p} | T = n] = K_p \sigma^{2p} n^{-p}, \qquad n \geqq 2,$$

$$R^*_{\sigma,c} = E_{\mu,\sigma}[K_p \sigma^{2p} T^{-p} + cT], \qquad \sigma, c > 0.$$

If the relation $\hat{\sigma}_n^2 = \sigma^2 \bar{W}_{n-1}$ is substituted into the definition of T, then T assumes the form $T = t + 1$, where

$$t = t_a = \inf\{n \geqq m - 1 : Z_n > a\}$$

with

$$Z_n = \left(\frac{n+1}{nl_{n+1}}\right) n \bar{W}_n^{-\alpha}, \qquad n \geqq 1.$$

Observe that the distribution of $W_1, W_2, \cdots$ is independent of μ and σ^2, so that the distribution of t_a depends only on a. The subscripts μ and σ^2 are omitted in probability statements concerning $W_1, W_2, \cdots$ and t_a. Next, suppose that l_n, $n \geqq 1$, are of the form

$$l_n = 1 + \frac{1}{n} l_0 + o\left(\frac{1}{n}\right) \quad \text{as } n \to \infty. \tag{10.2}$$

Then $(n+1)/nl_{n+1} = 1 + \Delta_n/n$, $n \geqq 1$, where $\Delta_n \to 1 - l_0$ as $n \to \infty$. Using this, one may show that the process Z_n, $n \geqq 1$, is of the form considered in Chapter 4. In fact, $Z_n = S_n + \xi_n$, $n \geqq 1$, where

$$S_n = n - \alpha n(\bar{W}_n - 1)$$

and

$$\xi_n = \tfrac{1}{2}\alpha(\alpha+1)\left(\frac{1}{U_n}\right)^{\alpha+2} n(\bar{W}_n - 1)^2\left[1 + \frac{1}{n}\Delta_n\right] + \Delta_n \bar{X}_n,$$

where $|U_n - 1| < |\bar{W}_n - 1|$ for $n \geqq 1$ and $X_i = 1 - \alpha(W_i - 1)$ for $i \geqq 1$. It is clear that S_n, $n \geqq 1$, is a random walk; it is easily seen that ξ_n, $n \geqq 1$, are slowly changing. See Example 4.1. Observe that the mean and variance of X_1 are $E(X_1) = 1$ and $D(X_1) = 2\alpha^2$. Observe also that $\xi_n \Rightarrow \alpha(\alpha+1)\chi_1^2 + (1 - l_0)$ as $n \to \infty$. In particular, $\xi_n/\sqrt{n} \to 0$ in probability as $n \to \infty$.

Lemmas 10.2–10.4 and Theorem 10.1 detail the relevant properties of t_a. It is assumed that (10.2) holds throughout. α is any positive number.

LEMMA 10.2. (i) $$\frac{t_a}{a} \to 1 \quad \textit{w.p. 1 as } a \to \infty;$$

(ii) $$t_a^* = \frac{(t_a - a)}{\sqrt{a}} \Rightarrow N[0, 2\alpha^2] \quad \textit{as } a \to \infty;$$

(iii) $$E\Big[\sup_{a>0} (a^{-1} t_a)^q\Big] < \infty \quad \textit{for all } q > 0;$$

(iv) $$\int_{t_a > 2a} t_a^2 \, dP \to 0 \quad \textit{as } a \to \infty.$$

Proof. Since $\xi_n/n \to 0$ w.p. 1, the first assertion follows directly from Lemma 4.1, and the second from Lemma 4.2. For the third, observe that $a^{-1} t_a \leqq l_{t_a} \bar{W}_{t_a - 1}^{\alpha}$ on $\{t \geqq m\}$ for all $a > 0$, so that

$$\sup_{a>0} a^{-1} t_a \leqq m + \sup_{n \geqq m} l_n \bar{W}_n^{\alpha}.$$

The third assertion then follows, since all powers of $\sup_{n\geqq 1} \bar{W}_n$ are integrable.

The proof of (iv) is similar to that of (4.9).

The next lemma shows that the probability of stopping early is strongly influenced by the initial sample size m.

LEMMA 10.3.

$$P\left\{t_a \leqq \frac{a}{2}\right\} \sim P\{t_a = m-1\} \sim C_{m-1} \cdot a^{-(m-1)/2\alpha} \quad \textit{as } a \to \infty,$$

where

$$C_n = \left[n\left(\frac{n+1}{l_{n+1}}\right)^{1/\alpha}\right]^{n/2} \Big/ 2^{n/2}\Gamma\left(\frac{n}{2}+1\right), \qquad n \geqq 1.$$

Proof. Let G_n denote the distribution function of χ_n^2 and let $C_n^* = 1/2^{n/2}\Gamma(n/2+1)$, $n \geqq 1$. Then $G_n(x) \leqq C_n^* x^{n/2}$ for all $x > 0$ and $G_n(x) \sim C_n^* x^{n/2}$

as $x \to 0$ for all $n \geqq 1$. Thus, as $a \to \infty$,

$$P\{t_a = m-1\} = P\{Z_{m-1} > a\}$$
$$= P\left\{(m-1)\bar{W}_{m-1} < (m-1)\left[\left(\frac{m}{l_m}\right)\frac{1}{a}\right]^{1/\alpha}\right\} \sim C_{m-1} \cdot a^{-(m-1)/2\alpha}.$$

In fact,

$$P\{t_a = n\} \leqq P\{Z_n > a\} \leqq C_n \cdot a^{-n/2\alpha}, \qquad n \geqq m-1, \quad a > 0$$

and

$$P\{t_a \leqq n\} = \sum_{k=m-1}^{n} P\{t_a = k\} \sim C_{m-1} \cdot a^{-(m-1)/2\alpha} \quad \text{as } a \to \infty,$$

for all $n \geqq m$. Next, using Stirling's formula, it is easy to see that there is a constant $b > 1$ for which $C_k \leqq b^k k^{k/2\alpha}$ for all $k \geqq 1$. So

$$P\{n < t_a < a^{3/4}\} \leqq \sum_{k=n+1}^{[a^{3/4}]} b^k \left(\frac{k}{a}\right)^{k/2\alpha} \leqq \sum_{k=n+1}^{\infty} b^k a^{-k/8\alpha},$$

which is of smaller order of magnitude than $a^{-(m-1)/2\alpha}$ as $a \to \infty$, provided that n is sufficiently large. Finally, $a^{3/4} \leqq t_a \leqq a/2$ implies that $Z_k > a$ for some $k \in (a^{3/4}, a/2]$ which implies that $\bar{W}_k^{\alpha} < (k+1)/al_{k+1}$ for some $k \in (a^{3/4}, a/2]$. For sufficiently large a, $(k+1)/al_{k+1}$ is bounded away from 1 from below for $a^{3/4} < k \leqq a/2$. So, there is an $\varepsilon > 0$ for which

$$\text{(10.3)} \qquad \begin{aligned} P\left\{a^{3/4} < t_a \leqq \frac{a}{2}\right\} &\leqq P\left\{\bar{W}_k - 1 < -\varepsilon, \exists k \in \left(a^{3/4}, \frac{a}{2}\right]\right\} \\ &\leqq P\left\{\max_{k \leqq a/2} k|\bar{W}_k - 1| > \varepsilon a^{3/4}\right\} \end{aligned}$$

for all sufficiently large a, and the last term in (10.3) is of smaller order of magnitude than $a^{-(m-1)/2\alpha}$ as $a \to \infty$, by the martingale inequality applied to a sufficiently high power of $k|\bar{W}_k - 1|$, $k \geqq 1$.

LEMMA 10.4. *If $m > 1 + 2\alpha$, then t_a^{*2}, $a \geqq 1$, are uniformly integrable.*

Proof. By Lemmas 10.2 and 10.3,

$$\int_{t_a \leqq a/2} t_a^{*2}\, dP + \int_{t_a > 2a} t_a^{*2}\, dP \to 0 \quad \text{as } a \to \infty,$$

so it suffices to show that there is a function J for which $xJ(x)$ is integrable with respect to Lebesgue measure over $(0, \infty)$ and

$$\text{(10.4)} \qquad P\left\{\frac{a}{2} < t_a < 2a, |t_a^*| > x\right\} \leqq J(x)$$

for all sufficiently large x and a. Indeed, the sufficiency of (4) follows by applying Theorem 1.5 to $t_a^* I_{\{a/2 < t < 2a\}}$.

Clearly, $P\{t_a > a/2,\, t_a^* < -x\} = 0$ for $x \geqq \sqrt{a}/2$ and, for $0 \leqq x < \sqrt{a}/2$, $t_a > a/2$ and $t_a^* < -x$ imply that $\bar{W}_k^\alpha < (k+1)/al_{k+1}$ for some $k \in (a/2, a - x\sqrt{a}]$, as in Lemma 3. For $a/2 < k < a - x\sqrt{a}$, sufficiently large x and sufficiently large a, one has

$$\frac{k+1}{al_{k+1}} \leqq \frac{[a+1-x\sqrt{a}-l_0+o(1)]}{a} \leqq 1 - \frac{x}{2\sqrt{a}},$$

so

$$P\left\{t_a > \frac{a}{2},\, t_a^* \leqq -x\right\} \leqq P\left\{\max_{k \leqq a} k|\bar{W}_k - 1| > \frac{x\sqrt{a}}{4}\right\},$$

which does not exceed Cx^{-4} for some constant C by the submartingale inequality applied to the submartingale $k^4(\bar{W}_k - 1)^4$, $k \geqq 1$. A similar, simpler argument shows that $P\{t_a < 2a,\, t_a^* > x\} \leqq Cx^{-4}$ for all sufficiently large x and a for some constant C to complete the proof.

In Theorems 10.1 and 10.2 below, ρ_α denotes the mean of the asymptotic distribution of the random walk S_n, $n \geqq 1$. Thus,

$$\rho_\alpha = \frac{1}{2} + \alpha^2 - \sum_{k=1}^{\infty} \frac{1}{k} E(S_k^-),$$

where S_k^- denotes the negative part of $S_k = k - \alpha k(\bar{W}_k - 1)$, $k \geqq 1$. Table 10.1 (p. 112) lists selected values of ρ_α.

THEOREM 10.1. *If $m > 1 + 2\alpha$, then*

$$E(t_a) = a + \rho_\alpha - \alpha(\alpha+1) + (l_0 - 1) + o(1),$$
$$E[(t_a - a)^2] \sim 2\alpha^2 a \quad \text{as } a \to \infty.$$

Proof. The first assertion follows from Theorem 4.5. The details of checking the conditions (4.10)–(4.16) are omitted; but note that (4.16) requires $m > 1 + 2\alpha$. The second assertion of the theorem follows immediately from Lemma 10.2(ii) and Lemma 10.4.

Observe that the regret $\mathcal{R}(\sigma, c)$ may be written

$$\mathcal{R}(\sigma, c) = acE\left[u\left(\frac{T}{a}\right) - u(1)\right], \qquad \sigma, c > 0,$$

where

$$u(x) = \frac{1}{p} x^{-p} + x, \qquad x > 0.$$

Thus, $\mathcal{R}(\sigma, c)/c$ is a function only of a.

THEOREM 10.2. *If $m > 1 + 2p$, then $\mathcal{R}(\sigma, c)/c \to p^2/(p+1)$ as $a \to \infty$.*

Proof. Since $u'(1) = 0$ and $u''(x) = (p+1)/x^{p+2}$,

$$u\left(\frac{T}{a}\right) - u(1) = \frac{(p+1)}{2}\left(\frac{1}{b}\right)^{p+2}\left(\frac{T}{a} - 1\right)^2$$

where b is an intermediate point with $|b-1|<|a^{-1}T-1|$. In particular, $b\to 1$ w.p. 1 as $a\to\infty$, and $b\geqq\frac{1}{2}$ on $T>a/2$ for all $a>0$. It follows that

$$\int_{T>a/2} a\left[u\left(\frac{T}{a}\right)-u(1)\right]dP=\frac{(p+1)}{2}\int_{T>a/2}\left(\frac{1}{b}\right)^{p+2}a^{-1}(T-a)^2\,dP$$

$$\to\frac{(p+1)2\alpha^2}{2}=\frac{p^2}{(p+1)}$$

as $a\to\infty$, by Theorem 10.1. Thus, it suffices to show that the integral over $T\leqq a/2$ tends to zero as $a\to\infty$. For $a\geqq 1$ and $T\leqq a/2$, $0\leqq u(T/a)-u(1)\leqq Ca^p$ for some constant C, so

$$0\leqq\int_{T\leqq a/2} a\left[u\left(\frac{T}{a}\right)-u(1)\right]dP\leqq Ca^{p+1}P\left\{T\leqq\frac{a}{2}\right\}=o(1)$$

as $a\to\infty$ by Lemma 10.3, since $m>1+2p$.

Remarks and references. The sequential procedure studied here was introduced by Robbins (1959) and initially developed by Starr (1966b), who noted the importance of the initial m. Relatives of Theorem 10.1 were discovered by Simons (1968) and Starr and Woodroofe (1968), (1969). Theorem 10.2 is adapted from Woodroofe (1977), but is implicit in Starr and Woodroofe (1969). For some extensions to nonnormal and multivariate contexts, see also Starr and Woodroofe (1972), Ghosh, Sinha and Mukhopadhyay (1976), Cabilio (1977) and Vardi (1979).

10.2. Fixed width confidence intervals. Again let $Y_1, Y_2, \cdots$ be i.i.d. normally distributed random variables with unknown mean μ, $-\infty<\mu<\infty$, and unknown variance $\sigma^2>0$. Now consider the problem of finding a confidence interval of width at most $2d$, $d>0$, and confidence coefficient γ, $0<\gamma<1$. Let $z=z(\gamma)$ be the $(1+\gamma)/2$ quantile of the standard normal distribution—that is, $\Phi(z)=(1+\gamma)/2$, where Φ denotes the standard normal distribution function. If σ^2 were known, then one could take a sample of size

$$n\geqq a=d^{-2}z^2\sigma^2,$$

and use

$$I_n=(\bar{Y}_n-d,\ \bar{Y}_n+d),$$

for

$$P_{\mu,\sigma}\{\mu\in I_n\}=P_{\mu,\sigma}\{|\bar{Y}_n-\mu|<d\}\geqq P_{\mu,\sigma}\left\{\frac{\sqrt{n}}{\sigma}|\bar{Y}_n-\mu|<z\right\}=\gamma.$$

However, if σ^2 is unknown, then it is impossible to find a confidence interval of width at most $2d$ and confidence coefficient $\gamma>0$, as demonstrated by Lehmann (1959, Problem 5.15).

For unknown σ^2, Anscombe (1953) suggested a sequential procedure which estimates σ^2 at each stage n and stops as soon as $n\geqq\hat{a}_n=d^{-2}z^2\hat{\sigma}_n^2$. As in § 10.1, we consider a simple modification: let $l_n \geqq 1$, $n \geqq 1$, and $l_n \to 1$ as $n \to \infty$, let $m \geqq 2$ be an initial sample size, let

$$T = \inf\{n \geqq m : n > d^{-2} z^2 l_n \hat{\sigma}_n^2\}$$

and consider the procedure which takes T observations and uses the interval $I_T = (\bar{Y}_T - d, \bar{Y}_T + d)$. Observe that T is of the form (10.1) with $\alpha = 1$ and $b = d^{-2} z^2$. Thus the distribution of T depends only on a.

The main result of this section gives an asymptotic expansion for $E(T)$ and the coverage probability $P_{\mu,\sigma}\{\mu \in I_T\}$. Using the independence of the event $\{T = n\}$ and $\bar{Y}_n$ for $n \geqq 2$, the latter may be written

$$P_{\mu,\sigma}\{\mu \in I_T\} = E\left[\psi\left(z^2 \frac{T}{a}\right)\right],$$

where $\psi(x) = 2\Phi(\sqrt{x}) - 1$, $x > 0$.

THEOREM 10.3. *Suppose that* (10.2) *holds. If* $m \geqq 4$, *then*

$$E(T) = a + \rho_1 + (l_0 - 2) + o(1) \quad \text{as } a \to \infty,$$

and if $m \geqq 7$ *then*

$$P_{\mu,\sigma}\{\mu \in I_T\} = \gamma + \frac{1}{a}[z^2 \psi'(z^2)(\rho_1 + l_0 - 2) + z^4 \psi''(z^2)] + o\left(\frac{1}{a}\right).$$

Proof. The first assertion follows directly from Theorem 10.1. To establish the second, expand ψ in a Taylor series about z^2 to find

$$P_{\mu,\sigma}\{\mu \in I_T\} = \gamma + \frac{1}{a} z^2 \psi'(z^2) E[T - a] + \frac{1}{2a} z^4 E\left[\psi''(W) \frac{1}{a}(T - a)^2\right],$$

where W denotes an intermediate value with $|z^2 - W| < z^2 |a^{-1} T - 1|$. Let $U_a = \psi''(W) a^{-1} (T - a)^2$, $a > 0$. Then the asymptotic distribution of U_a is $2\psi''(z^2)\chi_1^2$ as $a \to \infty$. Next, let $B = B_a$ be the event that $T > a/2$, $a > 0$. Then $\psi''(W)$ is bounded on B_a, so that $\lim E[U_a I_B] = 2\psi''(z^2)$ by Theorem 10.1. Moreover, there is a constant C for which $|\psi''(w)| \leqq C w^{-3/2}$ for all $w > 0$. Since $W > a^{-1} T z^2$ on $T \leqq a/2$, it follows that

$$\int_{B'_a} U_a \, dP \leqq C z^{-3} a \int_{B'_a} \left(\frac{a}{T}\right)^{3/2} dP \leqq C z^{-3} a^{5/2} P\left\{T \leqq \frac{a}{2}\right\}$$

which tends to zero as $a \to \infty$ by Lemma 10.3, since $m \geqq 7$. The second assertion of the theorem now follows.

As a corollary, we see that $P_{\mu,\sigma}\{\mu \in I_T\} > \gamma$ for all large a if

$$\rho_1 + l_0 > 2 - z^2 \frac{\psi''(z^2)}{\psi'(z^2)} = 2 + \frac{(1 + z^2)}{2}. \tag{10.5}$$

Observe that $\rho_1 \approx .82$ from Table 10.1.

Remarks and references. Sequential confidence intervals for the mean of a normal distribution were introduced by Stein (1945), (1949) and Anscombe

TABLE 10.1
Values of ρ_α

α	1/4	1/3	1/2	2/3	3/4	1
ρ_α	.549	.576	.633	.694	.724	.818

The computations were done on an Apple II microcomputer, using formula (26.4.6) of Abramowitz and Stegun (1970) to compute the chi-squared distribution function.

(1953). The techniques were extended to a nonparametric context by Chow and Robbins (1965). Starr (1966a) made an extensive study of the normal case. His computations indicate that (10.5) is not quite sufficient to make the coverage probability exceed γ for all a. Theorem 10.2 is adapted from Woodroofe (1977) and is closely related to results of Simons (1968) and Starr and Woodroofe (1968).

APPENDIX

Proof of the Renewal Theorem

For simplicity, the details are supplied only for the nonarithmetic case. Thus, let $X_1, X_2, \cdots$ be i.i.d. with common, nonarithmetic distribution F, and let S_n, $n \geqq 0$, denote the random walk $S_0 = 0$ and $S_n = X_1 + \cdots + X_n$, $n \geqq 1$. Suppose that F has a finite, positive mean μ and denote the characteristic function of F by

$$\phi(t) = \int_{-\infty}^{\infty} e^{itx} F\{dx\}, \qquad -\infty < t < \infty.$$

For $0 < s \leqq 1$, define the measure U_s on the Borel sets of $(-\infty, \infty)$ by

$$U_s\{B\} = \sum_{n=0}^{\infty} s^n P\{S_n \in B\}.$$

Then each U_s, $0 < s < 1$, is a finite measure, $U = U_1$ is the renewal measure and $U_s\{B\} \to U\{B\}$ as $s \uparrow 1$ for every Borel set B. Clearly, the Fourier transform of U_s is

$$\psi_s(t) = \frac{1}{1 - s\phi(t)}, \qquad -\infty < t < \infty, \quad 0 < s < 1.$$

Let

$$\psi(t) = \frac{1}{1 - \phi(t)}, \qquad t \neq 0.$$

The renewal theorem is proved by a careful analysis of the Fourier transforms ψ_s, $0 < s < 1$, and their limit ψ.

LEMMA A.1. *For any $c > 0$, $\int_0^c |\mathcal{R}\psi(t)|\, dt < \infty$, where $\mathcal{R}$ denotes real part.*

Proof. Since $\mathcal{R}\psi = (1 - \mathcal{R}\phi)/|1 - \phi|^2 \geqq 0$, the absolute value signs are superfluous. Now, for any $\varepsilon > 0$, $|1 - \phi|$ is bounded away from 0 on $[\varepsilon, c]$, since F is assumed to be nonarithmetic, so $\int_\varepsilon^c \mathcal{R}\psi(t)\, dt < \infty$. Next, for sufficiently small $\varepsilon > 0$, $|1 - \phi(t)| \geqq \mu t/2$ for $0 < t \leqq \varepsilon$, so it suffices to show that $t^{-2}[1 - \mathcal{R}\psi(t)]$ is integrable over $(0, \varepsilon)$ for small $\varepsilon > 0$. This follows easily from Fubini's theorem, since

$$\begin{aligned}
\int_0^\varepsilon t^{-2}[1 - \mathcal{R}\psi(t)]\, dt &= \int_0^\varepsilon t^{-2}\left\{\int_{-\infty}^{\infty} [1 - \cos(tx)] F\{dx\}\right\} dt \\
&= \int_{-\infty}^{\infty} \left\{\int_0^\varepsilon t^{-2}[1 - \cos(tx)]\, dt\right\} F\{dx\} \\
&= \int_{-\infty}^{\infty} \left\{\int_0^{\varepsilon|x|} t^{-2}[1 - \cos(t)]\, dt\right\} |x| F\{dx\} \\
&\leqq \left\{\int_0^{\infty} t^{-2}[1 - \cos(t)]\, dt\right\}\left\{\int_{-\infty}^{\infty} |x| F\{dx\}\right\} < \infty.
\end{aligned}$$

LEMMA A.2. *If g is any continuous function with compact support in* $(-\infty, \infty)$, *then*

$$\lim_{s\uparrow 1}\int_0^\infty g(t)\mathcal{R}\psi_s(t)\,dt = \frac{\pi}{2\mu}g(0) + \int_0^\infty g(t)\mathcal{R}\psi(t)\,dt.$$

Proof. Clearly, $\mathcal{R}\psi_s \to \mathcal{R}\psi$ boundedly on compact subintervals of $(0,\infty)$, so $\int_\varepsilon^\infty g\mathcal{R}\psi_s\,dt \to \int_\varepsilon^\infty g\mathcal{R}\psi\,dt$ as $s\uparrow 1$ for all $\varepsilon > 0$. Thus, it suffices to show that

$$\text{(A.1)} \qquad \lim_{\varepsilon\downarrow 0}\lim_{s\uparrow 1}\int_0^\varepsilon g(t)[\mathcal{R}\psi_s(t) - \mathcal{R}\psi(t)]\,dt = \frac{\pi}{2\mu}g(0).$$

Now,

$$\mathcal{R}\psi_s - \mathcal{R}\psi = \frac{-(1-s)}{|1-s\phi|^2}\mathcal{R}\left\{\frac{\phi(1-s\bar{\phi})}{1-\phi}\right\}$$

$$= \frac{-(1-s)}{|1-s\phi|^2}\mathcal{R}\left\{\frac{\phi(1-\bar{\phi})}{1-\phi}\right\} - \frac{(1-s)^2}{|1-s\phi|^2}|\phi|^2\mathcal{R}\psi = \Delta_1 + \Delta_2, \quad \text{say,}$$

where $\bar{}$ denotes complex conjugate. Clearly, $\int_0^\varepsilon \Delta_2\,dt \leqq \int_0^\varepsilon \mathcal{R}\psi\,dt$, which is independent of s and tends to zero as $\varepsilon\downarrow 0$. To estimate Δ_1, first choose $\varepsilon > 0$ so small that $\frac{1}{2}\mu t \leqq |1-\phi(t)| \leqq 2\mu t$ for $0 < t < \varepsilon$. Then $\frac{1}{2}\mu t \leqq |1-\bar{\phi}(t)| \leqq 2\mu t$ for $0 < t \leqq \varepsilon$, and $\phi(1-\bar{\phi})/(1-\phi)$ is bounded on $0 < t \leqq \varepsilon$. Next, make the change of variables $t = (1-s)w$. Then

$$\frac{1-s\phi(t)}{1-s} = 1 + \frac{s[1-\phi(t)]}{1-s} \to 1 + i\mu w \quad \text{as } s\uparrow 1,$$

and

$$\left|\frac{1-s\phi(t)}{1-s}\right|^2 > 1 + \frac{1}{8}\mu^2 w^2$$

for $0 < w < \varepsilon/(1-s)$ and $\frac{1}{2} < s < 1$. It follows easily that $(1-s)\Delta_1(t) \to 1/(1+\mu^2 w^2)$ for $w > 0$ as $s\uparrow 1$ and that $|(1-s)\Delta_1(t)|$ is dominated by some multiple of $1/(1+\mu^2 w^2)$ on $0 < w < \varepsilon/(1-s)$ for $\frac{1}{2} < s < 1$; it then follows from the dominated convergence theorem that

$$\int_0^\varepsilon g(t)\Delta_1(t)\,dt = \int_0^{\varepsilon/(1-s)} g(t)(1-s)\Delta_1(t)\,dw \to g(0)\int_0^\infty \frac{1}{1+\mu^2 w^2}\,dw = \frac{\pi}{2\mu}g(0)$$

to complete the proof of (A.1).

In the next lemma, $\mathscr{C}$ denotes the class of all integrable functions g whose Fourier transforms $\hat{g}(t) = \int_{-\infty}^\infty e^{itx}g(x)\,dx$ have compact support in $-\infty < t < \infty$. Observe that any such g may be written

$$g(x) = \frac{1}{2\pi}\int_{-\infty}^\infty e^{-itx}\hat{g}(t)\,dt, \qquad -\infty < x < \infty.$$

Observe also that any continuous, integrable function g is the limit of a sequence g_k, $k \geqq 1$, from $\mathscr{C}$ for, letting

$$h_k(x) = \frac{\sin^2(kx)}{\pi k x^2}, \qquad -\infty < x < \infty, \quad k \geqq 1,$$

one finds that the convolutions $g_k = g * h_k$ of any continuous integrable function g are in $\mathscr{C}$ for all $k \geqq 1$ and that g_k converge to g uniformly on compact subintervals of $(-\infty, \infty)$, as $k \to \infty$.

In the next lemma, let V_a denote the measure

$$V_a\{B\} = U\{a+B\} + U\{-a-B\}$$

for $a \geqq 0$. It will be shown that V_a converges weakly to $\mu^{-1} m_0$, where m_0 is Lebesgue measure, and that $U\{-a-J\} \to 0$ for all finite intervals J as $a \to \infty$.

LEMMA A.3. *If $g \in \mathscr{C}$ and $g \geqq 0$, then*

$$\lim_{a\to\infty} \int_{-\infty}^{\infty} g(x)\, V_a\{dx\} = \frac{1}{\mu} \int_{-\infty}^{\infty} g(x)\, dx.$$

Proof. Let $V_{s,a}\{B\} = U_s\{a+B\} + U_s\{-a-B\}$ for Borel sets $B \subset (-\infty, \infty)$, $a \geqq 0$, and $0 < s < 1$, and observe that $V_{s,a}\{B\} \to V_a\{B\}$ as $s \uparrow 1$ for all $a \geqq 0$ and B. Now, for $0 < s < 1$, the Fourier transform of $V_{s,a}$ is $2e^{-iat}\mathscr{R}\psi_s(t)$. So,

$$\int_{-\infty}^{\infty} g(x)\, V_{s,a}\{dx\} = \frac{1}{\pi} \int_0^{\infty} \hat{g}(t) 2\, e^{iat} \mathscr{R}\psi_s(t)\, dt$$

for all $g \in \mathscr{C}$ by Parseval's relation. Next, letting $s \uparrow 1$ and using Lemma A.2 yields

$$\text{(A.2)} \qquad \int_{-\infty}^{\infty} g(x)\, V_a\{dx\} = \frac{1}{\mu}\hat{g}(0) + \frac{1}{\pi} \int_0^{\infty} \hat{g}(t) 2\, e^{iat} \mathscr{R}\psi(t)\, dt$$

for nonnegative $g \in \mathscr{C}$. Finally, the integral on the right side of (A.2) tends to zero as $a \to \infty$, by the Riemann–Lebesgue lemma, and $\hat{g}(0) = \int_{-\infty}^{\infty} g(x)\, dx$ to complete the proof.

LEMMA A.4. *There is a $C > 0$ for which $V_a\{[-b, b]\} \leqq C(1+b)$ for all $a, b \geqq 0$.*

Proof. Given $b \geqq 0$, let $g = 4I_{[-b-1, b+1]}$, and let $h(x) = h_1(x) = \sin^2 x / \pi x^2$, $-\infty < x < \infty$. Then $g * h \in \mathscr{C}$ and $g * h \geqq I_{[-b,b]}$. So

$$\text{(A.3)} \quad V_a\{[-b,b]\} \leqq \int_{-\infty}^{\infty} g * h(x)\, V_a\{dx\} \leqq \frac{1}{\mu}\hat{g}(0)\hat{h}(0) + \frac{2}{\pi} \int_0^{\infty} |\hat{g}(t)\hat{h}(t)| \mathscr{R}\psi(t)\, dt,$$

by (A.2). Now, $|\hat{h}(t)| \leqq \hat{h}(0) = 1$ and $|\hat{g}(t)| \leqq \hat{g}(0) = 8(b+1)$ for all t, $-\infty < t < \infty$. so the first term on the right side of (A.3) does not exceed $8(b+1)/\mu$, and the second does not exceed $16\pi^{-1}(b+1) \int_0^{\infty} |\hat{h}(t)| \mathscr{R}\psi(t)\, dt$.

The four lemmas combine to prove the theorem as follows. First, it follows directly from Lemma A.4 that the family V_a, $a > 0$, is precompact. In particular, if a_k, $k \geqq 1$, is any sequence for which $a_k \to \infty$ as $k \to \infty$, then there are a subsequence k_j, $j \geqq 1$, and a measure V for which $V_a\{(b, c]\} \to V\{(b, c]\}$ as $a \to \infty$ along the subsequence a_{k_j} for all $b < c$ for which $V\{b\} = 0 = V\{c\}$. Let V be any limit point of V_a as $a \to \infty$. If $g \in \mathscr{C}$, $g \geqq 0$ and g has an integrable derivative g', then Lemma A.4 and integration by parts combine to show $\int g\, dV = \lim \int g\, dV_a$ as $a \to \infty$ along any sequence for which $V_a \to V$. Thus,

$$\int_{-\infty}^{\infty} g(x)V\{dx\} = \mu^{-1} \int_{-\infty}^{\infty} g(x)\, dx$$

for all such g, by Lemma A.3. It follows easily that $V = \mu^{-1} m_0$, where m_0 denotes Lebesgue measure. That is, V_a converges to $\mu^{-1} m_0$ as $a \to \infty$.

To complete the proof, it suffices to show that $U\{-a-J\} \to 0$ as $a \to \infty$ for all finite intervals J. There is no loss of generality in supposing that $J = [0, c]$, where $c > 0$. Let $\sigma_a = \inf\{n \geqq 1: S_n \leqq -a\}$. Then

$$U\{[-a-c, -a]\} \leqq P\{\sigma_a < \infty\} U\{[-c, c]\}, \qquad a \geqq 0. \tag{A.4}$$

Since $U\{[-c, c]\} < \infty$ by (A.2) and $P\{\sigma_a < \infty\} \to 0$ as $a \to \infty$ by the strong law of large numbers, the right side of (A.4) tends to zero as $a \to \infty$ to complete the proof.

Remarks and references. The proof of the renewal theorem is adapted from Feller and Orey (1961). See also Spitzer (1966) and Breiman (1968).

References

M. ABRAMOWITZ AND I. STEGUN [1970], *Handbook of Mathematical Functions*, National Bureau of Standards, Washington, DC.

F. ANSCOMBE [1952], *Large sample theory of sequential estimation*, Proc. Cambridge Philos. Soc., 48, pp. 600–607.

——— [1953], *Sequential estimation*, J. Roy. Statist. Soc., Ser. B, 15, pp. 1–21.

——— [1963], *Sequential medical trials*, J. Amer. Statist. Assoc., 38, pp. 365–83.

P. ARMITAGE [1957], *Restricted sequential procedures*, Biometrika, 44, pp. 9–26.

——— [1958], *Numerical studies in the sequential estimation of a binomial parameter*, Biometrika, 45, pp. 1–15.

——— [1963], *Comments on sequential medical trials*, J. Amer. Statist. Assoc. 58, pp. 383–387.

——— [1967], *Some developments in the theory and practice of sequential medical trials*, Proc. Fifth Berkeley Symposium Mathematical Statistics and Probability, 4, pp. 791–804.

——— [1975], *Sequential Medical Trials*, Halsted, New York.

P. ARMITAGE, C. K. MCPHERSON, AND B. C. ROWE [1969], *Repeated significance tests on accumulating data*, J. Roy. Statist. Soc., Ser. A, 132, pp. 235–244.

R. R. BAHADUR [1971], *Some Limit Theorems in Statistics*, CBMS Regional Conference Series in Applied Mathematics 4, Society for Industrial and Applied Mathematics, Philadelphia.

R. R. BAHADUR AND R. RAO [1960], *On deviations of the sample mean*, Ann. Math. Statist., 31, pp. 1015–1027.

O. BARNDORFF-NIELSEN [1978], *Information and Exponential Families in Statistical Theory*, John Wiley, New York.

L. BAUM AND M. KATZ [1965], *Convergence rates in the law of large numbers*, Trans. Amer. Math. Soc., 120, pp. 108–123.

G. BAXTER [1958], *An operator identity*, Pacific J. Math., 8, pp. 649–663.

P. BICKEL AND K. DOKSUM [1977], *Mathematical Statistics*, Holden-Day, San Francisco.

P. BILLINGSLEY [1968], *Convergence of Probability Measures*, John Wiley, New York.

D. BLACKWELL [1948], *A renewal theorem*, Duke Math. J., 15, pp. 145–150.

——— [1953], *Extension of a renewal theorem*, Pacific J. Math., 3, pp. 315–320.

D. BLACKWELL AND D. FREEDMAN [1964], *A remark on the coin tossing game*, Ann. Math. Statist., 35, pp. 1345–1347.

A. A. BOROVKOV AND R. A. ROGOZIN [1965], *On the multidimensional central limit theorem*, Theory Prob. Applic., 10, pp. 55–62.

L. BREIMAN [1968], *Probability*, Addison-Wesley, Reading, MA.

P. CABILIO [1977], *Sequential estimation in Bernoulli trials*, Ann. Statist., 5, pp. 342–356.

H. CHERNOFF [1972], *Sequential Analysis and Optimal Design*, CBMS Regional Conference Series in Applied Mathematics 8, Society for Industrial and Applied Mathematics, Philadelphia.

Y. S. CHOW AND H. ROBBINS [1965], *Asymptotic theory of fixed width confidence intervals for the mean*, Ann. Math. Statist., 36, pp. 457–462.

Y. S. CHOW, H. ROBBINS, AND H. TEICHER (1965), *Moments of randomly stopped sums*, Ann. Math. Statist., 36, pp. 789–799.

Y. S. CHOW, H. ROBBINS, AND D. SIEGMUND [1970], *Great Expectations*, Houghton Mifflin, Boston.

Y. S. CHOW, C. HSIUNG, AND T. LAI [1979], *Extended renewal theory and moment convergence in Anscombe's theorem*, Ann. Probab., 7, pp. 304–318.

K. L. CHUNG [1977], *A Course in Probability Theory*, 2nd ed., Academic Press, New York.

J. CORNFIELD AND S. GREENHOUSE [1967], *On certain aspects of sequential trials.* Proc. Fifth Berkeley Symposium Mathematical Statistics and Probability 4, pp. 813–829.

D. R. COX [1952], *A note on the sequential estimation of means,* Proc. Cambridge Philos. Soc., 48, pp. 45–450.

D. DARLING AND H. ROBBINS [1967a], *Iterated logarithm inequalities,* Proc. Nat. Acad. Sci., 57, pp. 1188–1192.

——— [1967b], *Inequalities for the sequence of sample means,* Proc. Nat. Acad. Sci., 57, pp. 1577–1580.

——— [1967c], *Confidence sequences for mean, variance, and median,* Proc. Nat. Acad. Sci., 58, pp. 66–68.

——— [1968], *Some further remarks on inequalities for sample sums,* Proc. Nat. Acad. Sci., 60, pp. 1175–1182.

B. EFRON [1978], *The geometry of exponential families,* Ann. Statist., 6, pp. 362–376.

P. ERDÖS, W. FELLER, AND H. POLLARD [1949], *A theorem on power series,* Bull Amer. Math. Soc., 55, pp. 201–204.

W. FELLER [1966], *An Introduction to Probability Theory and Its Applications,* Vol. 2, John Wiley, New York.

W. FELLER AND S. OREY [1961], *A renewal theorem,* J. Math. and Mech., 10, pp. 619–624.

M. GHOSH, B. SINHA, AND N. MUKHOPADHYAY [1976], *Multivariate sequential point estimation,* J. Multivar. Anal., 6, pp. 281–294.

C. HAGWOOD [1980], *Non-linear renewal theory for discrete random variables,* Comm. Statist., A9, pp. 1677–1698.

C. HAGWOOD AND M. WOODROOFE [1981], *Uniform integrability in non-linear renewal theory,* Ann. Probab., to appear.

R. KEENER [1980], *Renewal theory and the sequential design of experiments with two states of nature,* Comm. Statist., A9, pp. 1699–1726.

J. KIEFER AND L. WEISS [1957], *Some properties of generalized sequential probability ratio tests,* Ann. Math. Statist., 28, pp. 57–75.

T. L. LAI [1979], *Convergence rates and r-quick versions of the strong law for stationary mixing sequences,* Ann. Probab., 5, pp. 693–706.

T. L. LAI AND D. SIEGMUND [1977], *A non-linear renewal theory with applications to sequential analysis* I, Ann. Statist., 5, pp. 946–954.

——— [1979], *A non-linear renewal theory with applications to sequential analysis* II, Ann. Statist., 7, pp. 60–76.

S. LALLEY [1980], *Repeated likelihood ratio tests for curved exponential families,* Ph.D. thesis, Stanford University, Stanford, CA.

E. LEHMANN [1959], *Testing Statistical Hypotheses,* John Wiley, New York.

M. LOÈVE [1963], *Probability Theory,* Van Nostrand, New York.

G. LORDEN [1970], *On the excess over the boundary,* Ann. Math. Statist., 41, pp. 520–527.

——— [1973], *Open ended tests for Koopman Darmois families,* Ann. Statist., 1, pp. 633–643.

——— [1977], *Nearly optimal sequential tests for finitely many parameter values,* Ann. Statist., 5, pp. 1–21.

C. K. MCPHERSON AND P. ARMITAGE [1971], *Repeated significance tests on accumulating data when the null hypothesis is not true,* J. Roy. Statist. Soc. Ser. A, 134, pp. 15–26.

M. POLLAK AND D. SIEGMUND [1975], *Approximations to the expected sample size of certain sequential tests,* Ann. Statist., 3, pp. 1267–1282.

H. ROBBINS [1959], *Sequential estimation of the mean of a normal population,* in Probability and Statistics (the Harald Cramér volume), Almqvist and Wiksell, Stockholm.

——— [1970], *Statistical methods related to the law of the iterated logarithm,* Ann. Math. Statist., 41, pp. 1399–1410.

H. ROBBINS AND D. SIEGMUND [1970], *Boundary crossing probabilities for the Wiener process and sample sums,* Ann. Math. Statist., 41, pp. 1410–1429.

H. ROYDEN [1968], *Real Analysis,* Macmillan, New York.

G. SCHWARZ [1962], *Asymptotic shapes for Bayes sequential testing regions,* Ann. Math. Statist., 33, pp. 224–236.

——— [1968], *Asymptotic shapes for sequential testing of truncation parameters*, Ann. Math. Statist., 39, pp. 2038–2043.

H. SCHEFFE [1947], *A useful convergence theorem for probability distributions*, Ann. Math. Statist., 18, pp. 434–438.

D. SIEGMUND [1975], *Error probabilities and average sample number of the sequential probability ratio test*, J. Roy. Statist. Soc., Ser. B., 37, pp. 384–401.

——— [1976], *Importance sampling in the Monte Carlo study of sequential tests*, Ann. Statist., 4, pp. 673–684.

——— [1977], *Repeated significance tests for a normal mean*, Biometrika, 64, pp. 177–189.

——— [1978], *Estimation following sequential testing*, Biometrika, 65, pp. 341–349.

——— [1980], *Sequential* χ^2 *and F tests and the related confidence intervals*, Biometrika, 67, pp. 389–402.

D. SIEGMUND AND M. POLLAK [1975], *Approximation to the expected sample sizes of certain sequential tests*, Ann. Statist., 6, 1267–1282.

D. SIEGMUND AND P. GREGORY [1980], *A sequential clinical trial for testing* $p_1 = p_2$, Ann. Statist., 8, pp. 1219–1228.

G. SIMONS [1968], *On the cost of not knowing the variance when making a fixed width confidence for the mean*, Ann. Math. Statist., 39, pp. 1946–1952.

W. SMITH [1958], *Renewal theorem and its ramifications*, J. Roy. Statist. Soc., Ser. B., 20, pp. 243–302.

F. SPITZER [1960], *A Tauberian theorem and its probability interpretation*, Trans. Amer. Math. Soc., 94, pp. 150–169.

——— [1966], *Principles of a Random Walk*, Van Nostrand, New York.

N. STARR [1966a], *The performance of a sequential procedure for the fixed width interval estimation of the mean*, Ann. Math. Statist., 37, pp. 36–50.

——— (1966b), *On the asymptotic efficiency of a sequential procedure for estimating the mean*, Ann. Math. Statist., 37, pp. 1173–1185.

N. STARR AND M. WOODROOFE [1968], *Remarks on a stopping time*, Proc. Nat. Acad. Sci., 61, pp. 1215–1218.

——— [1969], *Remarks on sequential point estimation*, Proc. Nat. Acad. Sci., 63, pp. 285–288.

——— [1972], *Further remarks on sequential point estimation: the exponential case*, Ann. Math. Statist., 43, pp. 1147–1154.

C. STEIN [1945], *A two sample test for a linear hypothesis whose power is independent of the variance*, Ann. Math. Statist., 16, pp. 243–258.

——— [1946], *A note on cumulative sums*, Ann. Math. Statist., 17, pp. 489–499.

C. STONE [1965a], *Moment generating functions and renewal theory*, Ann. Math. Statist., 36, pp. 1298–1301.

——— [1965b], *On characteristic functions and renewal theory*, Trans. Amer. Math. Soc.

H. TAKAHASHI [1978], *On truncated power one tests and non-linear renewal theory*, Ph.D. thesis, Columbia University, New York.

H. TAKAHASHI AND M. WOODROOFE [1981], *Asymptotic expansions in non-linear renewal theory*, Comm. Statist., to appear.

Y. VARDI [1979], *Asymptotic optimal sequential estimation: the Poisson case*. Ann. Statist., 7, pp. 1040–1051.

B. VON BAHR [1965], *Convergence of moments in the central limit theorem*, Ann. Statist., 36, pp. 808–818.

A. WALD [1947], *Sequential Analysis*, John Wiley, New York.

M. WOODROOFE [1976a], *A renewal theorem for curved boundaries and moments of first passage times*, Ann. Probab., 4, pp. 67–80.

——— [1976b], *Frequentist properties of Bayesian sequential tests*, Biometrika, 63, pp. 101–110.

——— [1977], *Second order approximations for sequential point and interval estimation*, Ann. Statist., 5, pp. 984–995.

——— [1978], *Large deviations of the likelihood ratio statistic with applications to sequential testing*, Ann. Statist., 6, pp. 72–84.

——— [1979], *Repeated likelihood ratio tests*, Biometrika, 66, pp. 453–463.